Global Warming: Facts, Myths and Remedies

Joaquin Sitte

Global Warming: Facts, Myths and Remedies

A Primer

Springer

Joaquin Sitte
Waikanae, Wellington
New Zealand

ISBN 978-3-032-20861-3 ISBN 978-3-032-20862-0 (eBook)
https://doi.org/10.1007/978-3-032-20862-0

This Springer imprint is published by the registered company Springer Nature Switzerland AG
The registered company address is: Gewerbestrasse 11, 6330 Cham, Switzerland

To Renate
my dear wife

Preface

> ...if the quantity of carbonic acid[1] increases in geometric progression, the augmentation of the temperature will increase nearly in arithmetic progression.
> ...A simple calculation shows that the temperature in the Arctic regions would rise by about 8° to 9°C if carbonic acid increased to 2.5 or 3 times its present value.
>
> Svante Arrhenius[2] 1896

People had already shown interest in human-caused climate change in the eighteenth century (Fleming 1998). For example, David Hume concluded from ancient writings that the climate in Roman times must have been colder than the climate he was experiencing at the time (around 1750). Hume and other distinguished thinkers of the time attributed the mild climate they were experiencing to warming caused by land clearing. They argued that land clearing removed the ground's shielding from direct

[1] In Arrhenius' time CO_2 used to be called carbonic acid.

[2] Swedish physicist, 1903 Nobel Laureate in Chemistry.

sunlight provided by forests. There was also concern that the land cleared during the colonisation of North America might alter the climate there. The prevailing thinking at the time compelled Thomas Jefferson to call for the immediate start of climate measurements "before the climate has changed too much." Jefferson himself recorded the temperature twice daily for 50 years.

Recently, the preoccupation with *Climate Change* has returned to the spotlight after decades of warnings from scientists, and the ever more frequent news of extreme weather events and broken temperature records. The continuous stream of such news must wake up even the most complacent among us to the fact that the world's climate is warming.

The Intergovernmental Panel on Climate Change (IPCC), after decades of controversy between climate change deniers and climate researchers, finally asserted in its 2021 Sixth Assessment Report that: "It is unequivocal that human influence has warmed the atmosphere, ocean and land. Widespread and rapid changes in the atmosphere, ocean, cryosphere, and biosphere have occurred."

Still, there remain voices in the media that try to convince us that there is no evidence of human-caused global warming. Others do not deny global warming but argue that its consequences for most of the world's population are minor compared to other, more serious threats.

A common objection is that the current warming is merely part of the large variations in the Earth's climate over geological time, resulting from geological factors, variations in the Sun's radiation, and other factors. The *climate change critics* accuse the *climate scientists* of ignoring the geological past and holding the absurd view that the climate has been stable until suddenly the industrial age came along.

They accuse those who warn about the dangers of global warming, including the IPCC, of being alarmists and doomsday preachers. Undeniably, there is a great deal of alarmist rhetoric in the media. Nevertheless, the warming of our planet's climate is real and is of our own making. There is abundant open and trustworthy data that allows anyone to verify this conclusion.

Physics teaches us that, regardless of the complexity of systems, there are always first-order effects and second-order effects, as well as processes that unfold over different time scales. The current warming trend is on a historical timescale (hundreds of years, rather than a geological timescale of millions of years). When we examine observed variations in solar irradiance, it is clear that they are second-order fluctuations in total irradiance. We also have a quite sound model of how the small change in atmospheric greenhouse gas concentrations shifts the balance between incoming and outgoing radiation. In regulatory circuits, small changes in the controlling variable (greenhouse gases) cause large changes in the controlled variable (mean global surface temperature). The existence of this regulatory mechanism strongly suggests that the positive correlation between atmospheric CO_2 concentration and Mean Global Surface Temperature (GMST) is not merely a correlation but a causal relationship.

The warming of the atmosphere is a global phenomenon. Although each region of the globe has its own *climate*, conditioned by its latitude, topography, and its vicinity to the sea, the atmosphere covers every region with no regard for boundaries. Individual countries cannot isolate themselves from other countries' actions on the atmosphere, nor can they counteract Earth's climate warming with local actions alone. Therefore, any analysis of global warming requires a global perspective.

Based on what we know, we expect a warming of a few degrees centigrade over the next half-century. Such warming will not make us go extinct, but indications are that it will be **uncomfortable** for most of humankind and be detrimental to humanity's lifestyle, health, and wealth. Therefore, we should stop the warming and do it quickly.

For me, the question was: What do I need to know to understand the challenge of global warming? This book is my answer to this question.

In this book, I use public data and simple maths to show that anyone can verify **what** is happening, **why** it is happening, and **how** we can stop the warming. The media and political discourse often paint a confusing, overly complicated picture of global warming and energy technology. I wrote this book as a travel guide for motivated readers to acquire the fundamental knowledge they need to disentangle the confusion. The discussion around energy and global warming keeps changing. Therefore, to prevent us from losing sight of the essence of what is happening, I focus on the enduring fundamental relations.

The evidence for global warming has been there for quite some time for everyone to see. Records of the Earth's surface temperature date back to the mid-19th century and clearly show that the world's climate is warming at an alarming rate. The magnitude and speed of this global warming have already started to disrupt many people's lives, threatening to ravage how we sustain ourselves on this planet. Worldwide research has identified beyond a reasonable doubt that the primary cause of global warming is the massive release of carbon dioxide (CO_2) into the atmosphere. This CO_2 is the waste product from our burning of fossil fuels to obtain the energy that drives prosperity. We have created a technical civilisation that cannot exist without energy. As recently as 2024, over 80% of humankind's energy consumption came from burning fossil fuels. If we continue to obtain our energy from burning fossil fuels, the CO_2 emitted will severely degrade the civilisation we built on this practice. Therefore, changing how we generate energy is our urgent collective task. Any delay will greatly reduce the habitability of our planet.

The economic losses arising from severe weather events (droughts, floods, storms, and wildfires) are escalating (Ncei 2022) (Guo, Kubli, and Saner 2021). In the USA alone, over the 42-year period from 1980 to 2021, there were, on average, 7.7 severe weather events per year, each costing more than 1 billion (CPI-adjusted). This number increased to an average of 17.8 (!) events over the five years from 2017 to 2021. The estimated damage from the 20 events in 2021 was USD 145 billion.

Severe weather events cause production declines, which in turn reduce the availability of food and other essential supplies. Also, the consequential need for repairs increases demand in some sectors. By affecting supply and demand, severe weather events increase consumer prices, resulting in personal hardship and economic loss. The broader economic ramifications of severe weather events are more difficult to trace than direct insurance losses.

The Swiss Re Institute (Guo, Kubli, and Saner 2021) estimated that even if the world succeeds in limiting the temperature rise to the 1.5°C aimed for in the Paris Agreement, the percentage loss of world GDP by mid-century would be 4.2% (compared with no warming). The more likely scenario is that the temperature will rise between 2°C and 3.2°C, resulting in GDP losses estimated at 11% to 18%. There is no doubt that global warming will make us poorer.

The climate has been fairly stable over the Holocene (the last 12,000 years, during which civilizations emerged). There have been bursts of volcanic eruptions and other events, such as solar activity minima, that have disturbed the climate during the Holocene. Historian Peter Frankopan, in his book "The Earth Transformed" (Frankopan 2023), examines what we know about these changes and their impact on humankind. The effects have been profound, despite the mild climate change, compared to the changes that have occurred over geological time scales. History shows that our civilisation is fragile when it comes to climate variations. It is an illusion to think that, because of our new technological skills, we are immune to the consequences of these variations. Global warming, even when small on the geological scale, can wreak havoc on human society.

Global warming can have subtle consequences. For example, because of the more frequent and severe heat waves along the European Mediterranean coast, long-term annual European holiday makers are shifting their *summer holidays* to cooler parts of the year or are looking for cooler places to spend their summer holidays. Reductions as small as 10% in visitor numbers, as observed in 2023, can already put financial strain on tourism operators. As global warming intensifies, this trend is expected to continue.

Most people are now aware of global warming and the need to stop it. Sadly, decades of controversy over global warming and its causes have inhibited timely action to avoid its damaging consequences. The magnitude and speed of the warming demand immediate and decisive action to stop it. Lately, world decision-makers have flocked to international climate policy meetings, which may signal they have accepted the need for action. Sadly, no significant actions have followed, and instead, controversy persists about what actions to take and by whom, stifling effective action. Entrenched economic interests are keen to fuel this controversy in order to delay actions that threaten their business. The way the media reports about these issues amplifies the controversies, making it difficult for most people to judge the effectiveness of any proposed actions. Echoing the ill-informed views of the *environmentalist* lobby does not help. Many *environmentalist* groups blame the *fossil fuel industry* for global warming, ignoring that the fossil fuel industry thrives precisely because we consumers are eager to buy their products. The fossil fuel industry will be there as long as it finds customers. The "save the planet" slogan used by some extreme environmental groups is an exaggeration that exploits people's fears, making it harder for many to think clearly.

Instead of crying wolf, we need to think about effective actions. The grandiose actions to reduce our proverbial carbon footprint, promoted by the self-proclaimed *climate activist* to *save the world*, amount to **pussy-footing around the carbon footprint**. Cycling to work is undoubtedly beneficial for your health and helps reduce traffic congestion. Still, it will hardly stop global warming, even if it were a viable option for most commuters.

Paying for a carbon offset when you fly might make some people feel better, but it's akin to paying for forgiveness, allowing people to continue consuming fossil fuels. Flying less will help cut fuel consumption by a little. Small steps like these will not save the world. What these activists do not see is that it is not just about reducing carbon emissions; we need to stop using fossil fuels completely and quickly.

The fact that Hybrid Plug-in Electric Vehicles (HPEV) are so popular shows that, collectively, we are not prepared to sacrifice our lifestyles. The owners of HPEVs put their *comfort* of being able to travel farther between refills ahead of the need to eliminate fossil fuel consumption. An HPEV is a lesser electric vehicle (EV) combined with a lesser internal combustion engine (ICE) vehicle. The contribution of HPEVs to reducing global warming is demonstrably insignificant when considering the whole vehicle life cycle. Driving an HPEV may soothe the owner's environmental conscience, while manufacturers exploit this feeling to market their vehicles.

The concentration of CO_2 in the atmosphere continued to grow since the first United Nations global climate summit in Rio de Janeiro in 1992, despite the 29 Conferences of the Parties (COP) that followed. Under the United Nations Framework Convention on Climate Change (UNFCCC), global warming continues relentlessly. While we squabble about what to do, the entrenched interests in the status quo feed the paralysis.

It seems that world leaders will not take effective action to stop global warming until the warming has caused severe disruptions to humankind's way of life on this planet. The way to overcome this paralysis is to educate as many people as possible about how and why the warming is happening and recognise how to **reduce** global warming. Only the collective awareness arising from understanding our predicament can generate the social and economic momentum required for swift and decisive actions that do not deteriorate the comfortable life already achieved by part of humankind and do not prevent those still in poverty from achieving a dignified life.

Ultimately, *public opinion* is what will motivate political leaders to advocate for the necessary actions. And that public opinion must rest on facts and understanding of the processes involved in global warming and the technical means available to stop it.

It is actually easy to understand what causes global warming, and the key facts that show how it will affect our lives are available. Anyone interested can learn what they need to know to make informed decisions about climate change and energy.

Key facts about global warming, its causes, and effects are now readily available to everyone, thanks to the internet. Yet, even with the information readily accessible, finding the correct and relevant facts remains a time-consuming task for the uninitiated. Even more so is tying together the facts into a model of cause and effect. In this book, I have done some of this work for the reader. As humankind's journey unfolds toward a sustainable energy future, readers can update the data and keep pace with what is happening.

With the knowledge we have today and the technical capabilities we possess, the conclusion that emerges from the analysis is that transitioning to a fully electric economy powered by solar energy is the fastest and least painful way for us to mitigate global warming. Contrary to what some prominent authors (Gates 2022) claim, namely that much of the technology needed to achieve zero emissions either doesn't exist yet or is too expensive, such a transition is technically and financially feasible. We can stop global warming within a couple of decades without severe discomfort.

Waikanae, New Zealand Joaquin Sitte

Competing Interests The author has no competing interests to declare that are relevant to the content of this manuscript.

Contents

1 Introduction 1

2 How Big Is that Problem? 13

2.1 Prediction and the Meaning of Understanding 14

2.2 Science 15

2.2.1 Scientific Facts 16

2.2.2 Observation of the World 17

2.2.3 What Do Numbers Tell Us? 17

2.2.4 What Is the Mass of the Atmosphere? 20

2.2.5 Concentration of CO_2 in the Atmosphere 23

2.3 A Word About Energy and Its Units 26

2.3.1 Units of Energy 27

3 Life in the Greenhouse 29

3.1 Greenhouse Earth 31

3.2 Energy Balance in Space 32

3.3 Incoming Solar Energy 36

3.4 The Earth's Energy Balance 38

3.5 Greenhouse Gases 42
3.5.1 Infrared Absorption Spectra of Greenhouse Gases 43
3.5.2 Water Vapour 45
3.5.3 Carbon Dioxide 46
3.5.4 Methane 48
3.5.5 Ozone 51
3.5.6 Earth's Radiation Absorbed by the Atmosphere 51
3.6 Radiative Forcing 54

4 Global Warming 57
4.1 Our Current Predicament 58
4.2 Historical Global Temperature Data 59
4.2.1 The Global Mean Surface Temperature 60
4.3 Is CO_2 the Causes of the Temperature Rise? 62
4.3.1 The 800,000-Year Temperature Record 65
4.3.2 The 500 Million Year Climate Record 68
4.3.3 Changes in the Solar Irradiance 69
4.4 Temperature Rise Due to CO_2 72
4.4.1 Arrhenius' Calculations 73
4.4.2 Extrapolating the Recent Trend in GMST Rise 76
4.5 The CO_2 Emission Budget 78
4.5.1 Natural CO_2 Removal 80
4.5.2 The Revised CO_2 Emission Budget 82
4.6 Contribution to Warming by Each Greenhouse Gas 84
4.6.1 The Methane Contribution 87

4.6.2 Human and Animal Respiration 92
4.7 The Weather 92
4.8 The Bottom Line 93

5 Fossil Fuels 95
5.1 CO_2 Produced by Fossil Fuel Combustion 96
5.1.1 Methane 96
5.1.2 Liquid Fossil Fuels 100
5.1.3 Coal 102
5.2 Specific Energy of Fuels 103
5.3 Global CO_2 Emission from Burning Fossil Fuels 106
5.4 Fossil Fuels, the Military and Conflict 108

6 Addicted to Energy 109
6.1 World Energy Consumption 110
6.1.1 When Energy Is *Consumed, Where Does It Go?* 115
6.2 Per Capita Energy Consumption 116
6.2.1 What Counts Towards Per Capita Energy Consumption? 118
6.2.2 What Can One Do with 72 GJ? 119
6.3 Energy Consumption and Wealth 120
6.4 Income and Happiness 124
6.5 The Inequality in Energy Consumption 125
6.6 Energy Consumption and International Trade 127
6.7 Energy Consumption Targets 129
6.8 Energy Waste 130

7 Alternatives to Fossil Fuels 133
7.1 The Hydrogen Economy 134
7.1.1 Hydrogen, Some Basic Facts 134
7.1.2 Where Do We Find Hydrogen? 136
7.1.3 Example: Driving on Hydrogen 137
7.2 Insisting on Hydrogen 140
7.2.1 Hydrogen Derivative Fuels 143
7.3 Nuclear Fission Power 147
7.4 Nuclear Fusion Power 148
7.5 CO_2 Capture 149
7.6 Solar Energy 150
7.7 Geothermal Energy 153
7.8 Biomass 154
7.9 End-User Energy from Fossil Fuels 155
7.10 Zero Emission Life-Cycle 159
7.11 The Bottom Line 159

8 Solar Power 161
8.1 Thermal Solar Energy 162
8.2 Photovoltaic Solar Electricity 162
8.2.1 Energy Generation Capacity of Solar Panels 163
8.3 PV Solar Energy Land Requirement 168
8.4 Solar Panel Cost 169
8.4.1 Making Photovoltaic Solar Cells 172
8.4.2 Fabrication Energy 173
8.4.3 CO_2 Emission Cost 177
8.4.4 Production Capacity 177
8.4.5 Metallurgical Grade Silicon Supply 180
8.4.6 Effect of Solar Panels on the Radiation Balance 181

8.5 Wind 183
8.5.1 Energy Generation Capacity of Wind Turbines 184
8.6 Storing Solar Energy 192
8.6.1 Batteries 193
8.6.2 Gravity Storage 200
8.6.3 Thermal Energy Storage 201
8.7 Solar Power Plants in Outer Space 203

9 Transition to an All Electric Economy 205
9.1 Paying for the Transition to Electricity 206
9.1.1 Carbon Emission Fee 206
9.1.2 Policies for Carbon Pricing 209
9.1.3 The Tax on CO_2 213
9.1.4 The Reduction of Fossil Fuel Consumption 214
9.1.5 Optimistic Guess of Future Temperature Rise 216
9.1.6 Impact on Panel Production Capacity 217
9.1.7 Implications of the Policy 218
9.1.8 How to Collect the Carbon Pricing Revenue 219
9.2 Electric Cars 219
9.2.1 Vehicle Emissions 220
9.2.2 Money Savings for the EV Owner 225
9.2.3 EV Purchase Subsidies 226
9.2.4 Hybrid Vehicles 227
9.2.5 The CO_2 Emission Cost of Car Manufacture 228
9.3 Air Travel 229

10 The Demise of Fossil Fuels 233
10.1 Fossil Fuel Industry Contribution to World GDP 234
10.2 Geopolitical Consequences 238
10.2.1 The Global Fossil Fuel Market 239
10.2.2 USA 241
10.2.3 Russia 241
10.2.4 Saudi Arabia 242
10.2.5 Bangladesh 243
10.2.6 India 244
10.2.7 China 245
10.2.8 The Bottom Line 246
10.3 An Example: Germany's Plan 246
10.4 Demand for Critical Minerals for the All-Electric Economy 248
10.5 Global Warming and Population Growth 249
10.6 Consequences of Adaptation 249

Epilogue 251

Appendix A: Baetz' Law 253

References 257

Index 263

About the Author

Joaquin Sitte is a researcher and retired academic. One of his passions is to reveal the underlying simplicity of apparently complex phenomena. He has worked on three continents and within diverse cultures. Trained as a physicist at the Universidad Central de Venezuela he holds a doctorate in Quantum Chemistry from Uppsala University in Sweden.

He spent the first half of his academic career in Venezuela founding and leading a research Group on the physical chemistry of solid surfaces. Together with his engineer wife and two children he resettled in Australia in the mid 1980s where he took up an academic position in Information Technology. During his career he has published over a hundred peer reviewed research papers in physics, computational intelligence and robotics.

1

Introduction

The better-off part of the world's population enjoys a comfortable lifestyle thanks to abundant, cheap energy. For as long as we can remember, most of us have not really thought about how easily we get energy, except when higher fuel or electricity prices suddenly remind us how much our comfort depends on having plenty of energy.

At the beginning of the 21st century, the observed warming of the Earth gained widespread public attention. Soon after, it also came to public attention that the atmospheric carbon dioxide concentration had risen steadily since the Industrial Revolution began in the 18th century. Scientists had been putting two and two together for quite some time. In 1896, Nobel prize winner Svante Arrhenius published his calculations of how much the global temperature would rise due to increasing the amount of carbon dioxide (CO_2) in the atmosphere. In the early 1980s, fossil fuel producers began to wonder what effect the carbon dioxide emissions from burning fossil fuels would have on the environment. An internal study by Shell Petroleum (Jacobs 1988) clearly

J. Sitte, *Global Warming: Facts, Myths and Remedies*,
https://doi.org/10.1007/978-3-032-20862-0_1

describes the situation. Yet, it took until 2021 for the Intergovernmental Panel on Climate Change to assert in its Sixth Assessment Report that "It is unequivocal that human influence has warmed the atmosphere, ocean and land."

Because the increase in the amount of CO_2 in the atmosphere comes from the burning of fossil fuels to provide the energy to sustain our industrial civilisation, it stands to reason to attribute the observed warming of the climate to humankind's consumption of fossil fuels. Such a hypothesis has severe implications for humankind's future well-being. Therefore, we need to take a close look at the supporting facts.

Measurements of Earth's surface temperature worldwide over nearly two centuries indicate that the Earth is warming. Figure 1.1 shows the Mean Global Surface Temperature for October, from 1850 to 2024, taken from the US National Oceanic and Atmospheric Administration (NOAA) global mean surface-temperature dataset , version 6.0 Huang et al. (2024).[1] I chose October because that is when temperatures are somewhere between the summer high and the winter low in both the northern and southern hemispheres.

The data show that the global mean surface temperature in October rose from 13.51 °C in 1850 to 14.97°C in 2024.

The global mean surface temperature varies slightly due to inherent temperature fluctuations and uncertainties in the measurement process. In Chap. 4, we will examine the data more closely.

Besides measuring temperature directly, we can also estimate what the Earth's surface temperature was like millions of years ago. Some ice cores enable us to estimate temperatures dating back up to 3 million years ago. By drilling into the ocean floor sediments, we can get clues about

[1] I have purposefully chosen to plot the data on a scale starting from zero to show that despite the change being small in relation the seasonal variation in surface temperatures, rising trend is clearly visible.

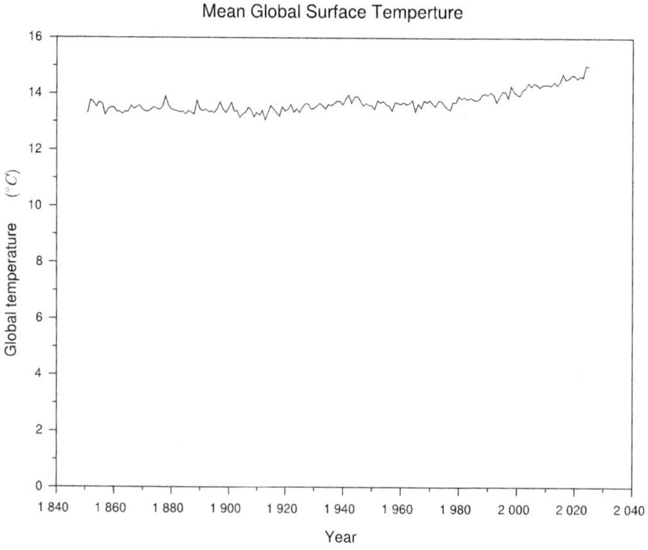

Fig. 1.1 Global mean surface temperatures for October from 1850 to 2024 (Data from NOAA)

temperatures from hundreds of millions of years ago. These records show that Earth has had times when it was over 10°C hotter than today.

Some authors use this to belittle the temperature rise in Fig. 1.1. However, the fact that the Earth has experienced long-term warm periods in the distant past is irrelevant to the warming observed over the last couple of hundred years. All life on Earth has thrived in the climate of the last 10,000 years, since the end of the last ice age. What is important to us is how a new warming era will affect the existing biosphere.

A rising mean surface temperature does not necessarily mean that the warming is uniform across the globe, nor does it imply that every year is warmer than the previous year. The world is a large space where land and sea cover vast expanses, where local changes can overshadow long-term trends. For this reason, some writers prefer the term climate change over global warming. Despite the uneven nature of regional climate change, the long-term trend is clear: the global climate is warming. Global warming is the problem, and we should not obfuscate the issue by ambiguously calling it climate *change.*

Figure 1.2 shows the carbon dioxide (CO_2) concentration in the atmosphere over the last eight hundred thousand years (EPA 2022). We can know the values for the atmospheric CO_2 concentration that far back in time, thanks to tiny air bubbles trapped in the layers of Antarctic ice. The data show that, since around 800,000 years ago until the last hundred years or so, the CO_2 concentration, despite rising and falling in cycles, has remained between 200 and 300 parts per million (ppm). Since then, the CO_2 concentration has risen sharply, surpassing 400 ppm. Regular direct measurements of the atmospheric CO_2 concentration only started in 1958 at the Mauna Loa Observatory in Hawaii by Charles David Keeling (Keeling 1960). These measurements continue to this day, and the data, shown in Fig. 1.3, became known in the literature as the*Keeling Curve.* The curve expands the sharp peak on the right-hand end of the 800,000-year record in Fig. 1.2. When comparing the Keeling curve with the surface temperature record, we see that the rise in global mean surface temperature goes hand in hand with the increase of the CO_2 concentration in the atmosphere (see Fig. 4.3). Since Arrhenius' pioneering calculations, we know this is not just a correlation but a cause-and-effect relationship. Through the *Greenhouse Effect*, the amount of CO_2 in the atmosphere regulates the global surface temperature. In Chap. 3, we will analyse in detail how this works.

As Fig. 1.2 shows, CO_2 has been in the air for a long time and is essential for life on Earth. But CO_2 is also the waste we release into the air when we get energy by burning coal,

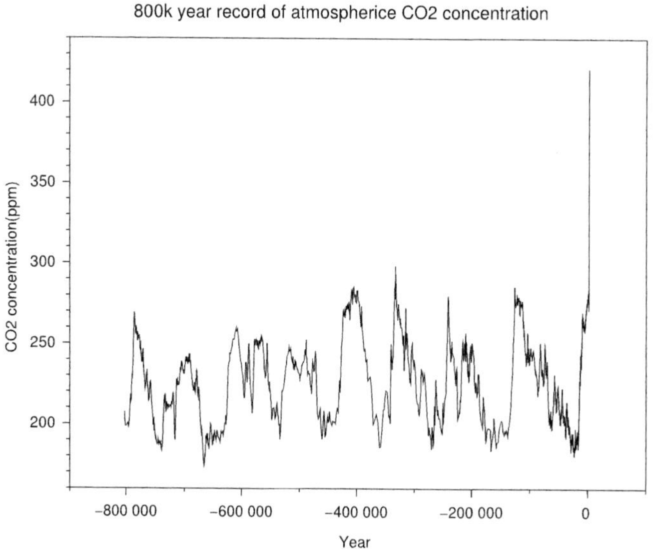

Fig. 1.2 Atmospheric Carbon dioxide concentration over the last 800 thousand years (Data from U.S. EPA)

oil, or gas. Coal, oil, and gas are types of carbon compounds found underground. We call them *fossil fuels* because they come from old plant and animal remains buried for millions of years. This extra CO_2 we release into the atmosphere is the reason for the recent *abnormal* rise in CO_2 levels shown in Fig. 1.2. We also release other gases that alter the Earth's temperature, but CO_2 has the greatest effect.

There we have it: We are warming the planet by burning fossil fuels to supply the energy craved by our lifestyle. But does it matter?

The temperature record in Fig. 1.1 shows that by 2024, the temperature rose by 1.46 °C since 1850. A temperature rise of one and a half degrees may not seem much. Inhabitants of colder climate zones may even welcome a bit of warming. Unfortunately, we already know that small changes

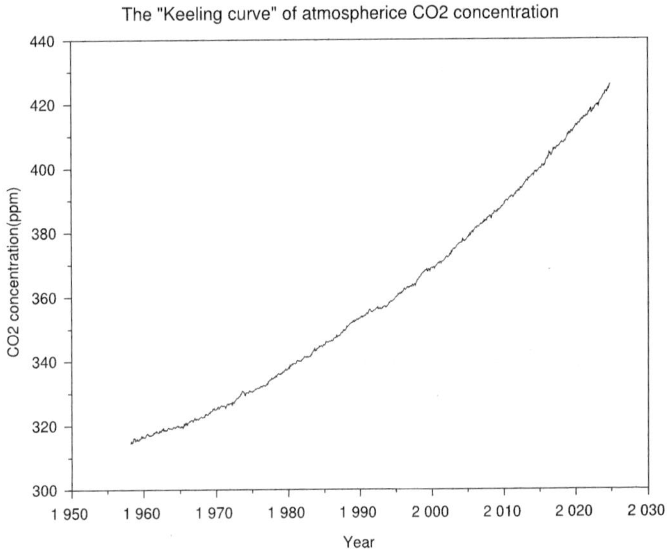

Fig. 1.3 The *Keeling Curve* of atmospheric Carbon Dioxide concentration, seasonally adjusted, since 1958

in average temperatures can have a big effect on life on Earth. Geological records show the Earth has undergone various heating and cooling periods (see Fig. 4.4). Even though these temperature variations have played out over long periods, the consequences of these changes for life on Earth have been profound. As for today, the effects of the small increase of 1.5 °C over the last hundred years are already being felt and are a matter of concern.

Glaciers are shrinking. Polar ice caps are melting, particularly in the Arctic. Sea levels are rising. Extreme Weather is becoming more severe and frequent. We must remember that natural ecosystems have adapted over millennia to flourish under prevailing conditions. Our climate has been relatively stable since the end of the last ice age 12,000 years ago. The recent warming trend threatens the habitat of many plant and animal species and the livelihoods of significant portions of the World's human population. Only two degrees of warming can almost entirely prevent sea ice formation in the Arctic over most of the year. Polar bears need extensive sea ice to hunt their primary food source: seals. Likewise, farming of many crops depends on yearly repeating weather patterns. Their disruption by even modest changes in rainfall and hot spells can make cultivation difficult, if not impossible, thereby erasing the livelihoods of millions of people. Humankind may adapt to the changes, and so may many ecosystems, but it is clear that at just a few degrees warmer, the World will be very different. Indications are that it is not something to look forward to. Besides, we do not know what else will happen when the temperature continues to rise.

The mean surface temperature over the 100 years from 1901 to 2000 was 13.9 °C. Extending the trend of temperature over the last 40 years to 2050 yields a mean temperature rise of 1.31°C above the 100-year mean and 2.37°C by 2100 (see Fig. 4.1). Detailed climate models predict even bigger rises of 3°C or more if we do not reduce burning fossil fuels.

We must conclude that generating energy from fossil fuels now poses a threat to the very prosperity that fossil fuels provide.

Our first reaction might be: Let's eliminate fossil fuels! Indeed, that is what we have to do. However, it is easier said than done.

On the positive side, most of our leaders now admit that global warming is caused by burning lots of fossil fuels and that it is a serious threat to humankind. They even meet at international conferences to collaborate on solving the problem. On the negative side, the actions they have agreed to are not enough. Worse still, they often fail to follow through. Leaders put off dealing with the problem. It seems that people find it hard to remedy the situation. We can only guess why there is so little action. Is it because leaders do not understand the problem? Is it because they cannot agree? Is it because they are not motivated? Or is it just human nature to ignore threats that do not seem urgent? Perhaps all of these reasons play a part. Merz et al. (Merz 2023) argue that we are in a behavioural crisis driven by addiction to economic growth, dominated by marketing, and a superseded pro-natalist mentality that results in an overshoot of consumption, waste, and population. It is the intentional exploitation of previously adaptive human impulses that promotes and perpetuates maladaptive behavior, leading to overshoots.

One thing is clear, though: humanity's current prosperity relies on abundant energy, and we want to continue enjoying the lifestyle it provides. Therefore, actions to stop global warming will only receive widespread public support if they maintain a high-energy-consuming lifestyle. This dilemma will shape the first half of the 21st century. Solving this dilemma will push our social and technical skills to the limit. The more people understand the magnitude and nature of the task, the easier it will be to solve.

We need to know: What sources of energy can replace fossil fuels? How much energy can we obtain from each of the alternative sources? Can we install sufficient generation capacity to replace fossil fuels before the climate gets too hot? What will the energy cost? What are the negative side effects of exploiting promising alternative energy sources? What are the social and economic consequences of replacing fossil fuels in a short time frame? Can a proposed energy source fulfil the promises made by its promoters? Can we reverse warming by extracting carbon dioxide from the atmosphere in sufficient quantity at an affordable cost? How much must we drive an electric vehicle to offset the carbon debt incurred during its manufacture? The answers to these questions, and many more that will arise in the future, will enable us to make informed decisions and lifestyle choices as consumers, workers, business managers, investors, or citizens at the ballot box.

This book aims to analyse, in the most transparent way, the driving forces of global warming, our energy use, and the state of available energy generation technologies. What we learn from this analysis will help us identify actions we can take and estimate their consequences. Let us not be naive or easily manipulated by our ignorance. Let us shine a bright light on the issue to dispel myths and misunderstandings, revealing the situation in sharp clarity.

This book consists of three parts, each dedicated to one of the key issues:

- How is it that we have a *liveable* climate on Earth, and what regulates it?
- How much energy does our affluent lifestyle consume, and how does our energy consumption shift the climate?
- What can we do to replace fossil fuels with other sources of energy that do not disturb the climate?
- Can we muster the resources required to bring about the replacement of fossil fuels?

The answers to these questions must rest on numbers. We need numbers to separate the less relevant from the most important. To make such judgements, we need values for the various intervening quantities. Since we are dealing with the entire world, these values are rather large. For example, the amount of CO_2 emitted by the global population in 2019 was 36.44 Gt. In our daily life, we usually do not have to deal with such big masses, and therefore, most of us will not be able to attach, straight away, a *meaning* to such a large number. As preparation for some large-number manipulation, we start with a *mental training session* in Chap. 2.

Once we have developed some familiarity with the most relevant large quantities, we begin by asking, in Chap. 3, what makes our planet livable for us? The answer is the energy that our planet receives from the sun. Our climate depends on the energy balance between the incoming energy from the sun and the energy returned into space. Understanding the various factors that affect this energy balance is crucial to our ability to influence the climate.

Armed with this knowledge, we embark in Chap. 4 on estimating the future temperature rises our fossil fuel energy consumption may cause. This leads us to the notion of a carbon (CO_2) *budget* that tells us how much more CO_2 we can dump into the atmosphere before the temperature rise reaches a set limit.

Once we find the limit for CO_2, in Chap. 5 we translate that into limits on fossil fuel consumption and, consequently, into allowable fossil fuel energy generation. To understand what that means for our prosperity, we examine in Chap. 6 the intimate coupling between our prosperity and our energy consumption. The question of what options we have for sustaining our energy consumption, and therewith our prosperity, without burning fossil fuels is the subject of Chap. 7. This survey leads us to conclude that the most effective way to limit global warming to a tolerable level while maintaining our energy consumption is to replace fossil fuels with solar energy (photovoltaic solar panels, wind turbines, and hydroelectric dams). Therefore, in Chap. 8, we brush up on the technicalities of photovoltaic solar and wind energy generation to assess the viability of replacing fossil fuel energy with solar energy in the shortest possible time. How to finance such a transition to an all *electric economy* is the subject of Chap. 9. A transition to an electric global economy will break the current fossil-fuel world economy's dependence on a small number of fossil-fuel-producing nations. Speculation on the geopolitical consequences of the demise of the fossil fuel economy is the subject of the final Chap. 10.

2

How Big Is that Problem?

If we believe the media, on one side, global warming poses a threat to our prosperity and even to our existence, while on the other side, it is a hoax and a doomsday prophecy. How can we determine the true extent of the menace of global warming? To answer this question, we need to assign numerical values to the various factors at play in the problem, allowing us to make comparisons. Furthermore, to apply remedies against the harmful effects of global warming, we need to be able to make quantitative *predictions* of the effects our actions will have on the global warming process. If we cannot predict the effect of our actions, we will be unable to choose the action that produces the right outcome, and we will do nothing more than bet on a lottery.

The purpose of this chapter is to hone our skills in working with numbers to make simple *quantitative models* of real-world processes. For this, we need to be able to attach meaning to numbers and find relations between them. Throughout the book, we will use these skills to first relate our energy consumption with our CO_2 emissions and the mean surface

J. Sitte, *Global Warming: Facts, Myths and Remedies*,
https://doi.org/10.1007/978-3-032-20862-0_2

temperature of the Earth, and second, to gauge the effectiveness, viability, and affordability of alternative energy generation technologies. But first, let us review of the fundamental notions that underpin successful prediction.

2.1 Prediction and the Meaning of Understanding

The simple fact is that creatures that can correctly predict what will happen next have a better chance of survival. The hunter who correctly predicts the fleeing prey's path has a better chance of intercepting it and making a kill. Therefore, humans are inherently curious and seek to *explain* what happens in the world. If we can explain a situation, then we *understand* it and can make correct predictions about what will happen. But, what does *explaining* mean?

When we explain a situation, we describe how we think the various elements relate. This mental picture constitutes our *model* of the situation and reflects our *understanding* of the situation. The model enables us to mentally preview possible events and predict their outcomes.

Thus, the essence of *understanding* is creating a model that allows us to predict the outcome of some happening. The throwing of a spear needs a model that relates the spear's trajectory to the arm's throwing motion. Models are built on experience. When throwing a spear or hitting a tennis ball with a racket, the model is subconscious. We repeatedly practice hitting a target until we execute the throwing motion that makes the spear hit the target. Each time we throw the spear, we test our model. If the spear misses the target, we try to make a better throw next time; that is, we try to improve our model. Our subconscious model must be quantitative, even if we are not fully aware of how our brain does it. The target is at a certain distance, our throwing motion must follow a precise trajectory, and we need to exert a specific force.

For situations that do not involve sensory-motor control of our own body, we create conscious mental models in which we relate the observed magnitudes. For example, with the invention of cannons, soldiers needed a model that related the cannon's position, orientation, and inclination, along with the amount of gunpowder, to the target's location. Newton's equations of motion finally provided the model to predict where the cannonball, or today's ballistic missile, would hit. If we know how to use Newton's model (equations), we can claim to *understand* how ballistic missiles and other bodies move. In making models, it helps to keep in mind the *Occam's razor* principle of not making a model more complicated than needed.

2.2 Science

Over the last few centuries, the search for explanations (model building) of what happens in the world around us has been pursued ever more intensely, leading to what we call *Science*. Scientific *research* involves carefully observing the world around us and formulating models that explain the observations. When a model produces a wrong prediction, we conclude that it is faulty and needs to be modified or replaced with a new model. The improved or new model must lead to correct predictions in situations where the old model failed, while still providing correct predictions across the range of conditions where our previous model was correct. The *scientific method* prescribes how to conduct scientific research. The scientific method consists of a sequence of steps that start with formulating a hypothesis, which, in our vocabulary, is a model that is then subject to tests (predictions). If the tests confirm the hypothesis (i.e., the predictions are correct), then our confidence in its validity increases. Whereas a negative test result *falsifies* the hypothesis. A collection of hypotheses concerning a situation or a family of related situations is often referred to as a theory. In our terminology, a theory is just a model. A falsification may set boundaries to the range of conditions for which the theory is valid, and depending on the formulation of the hypothesis, it may even render the theory invalid altogether. A falsified hypothesis calls for a modified or new hypothesis that, hopefully, explains the observations.

For example, Newton's model for predicting the motion of material bodies yields accurate predictions when applied to objects of the sizes we encounter in our daily lives. It is accurate enough to, for example, provide the path for a spacecraft to orbit one of Jupiter's moons. However, Newton's model yields incorrect predictions when applied to electrons orbiting an atomic nucleus. Newton's model is useless for explaining the observations about the structure of matter, where objects are as small as atoms and elementary particles. To explain the observations about the structure of matter, we needed the new model of quantum mechanics.

2.2.1 Scientific Facts

Scientists conduct scientific research to explain situations or phenomena they do not yet understand. They may find that existing models explain the observations, or they may propose an extension to an existing model. Either way, the scientist submit their results in a research paper for publication in one of the peer-reviewed journals that cover their scientific discipline. Then, several, often three, reputable researchers in the same discipline (peers) review the paper. Only if the reviewers are satisfied that the paper's authors have flawlessly followed the scientific method and the paper meets other presentation standards will the paper be published. Once published, the paper becomes part of the public scientific knowledge and a trusted *scientific fact* until other researchers find flaws or further research invalidates the authors' model.

Peer review is not perfect. Faulty papers can occasionally slip through the peer review process. Even so, it is the best quality control for scientific research that we have and can afford. Peer-reviewed scientific research is the most trustworthy source of knowledge we have.

2.2.2 Observation of the World

Measurement is the foundation for our understanding of the world around us. Measurement means assigning numerical values to the quantities we observe. Expressing observed quantities as tiny, small, big, and huge is not enough. Unless our reasoning is grounded in numbers, all our conclusions will remain hazy. Unless we can put a number on the amount of carbon dioxide emitted globally in 2018 and 2019, we cannot say whether we emitted more or less in 2019 than in 2018. And, if it increased, how much did it increase? Is the increase negligible, so we do not need to worry about it, or is it a concern? We can only draw useful conclusions and decide on practical actions when we have numerical values for the observed quantities. Accordingly, our models must be capable of making quantitative predictions.

2.2.3 What Do Numbers Tell Us?

Data collectors tell us that in 2019, the world population emitted 37.04 Gt of carbon dioxide as a waste product from the combustion of fossil fuels (Olivier and Peters 2020). That number may not mean much to us except that it seems *large*

(Anything that is Giga is big!). When I read that number, questions immediately pop up in my mind: What does it measure? How large is it? How can I compare the number with another one that already has meaning for me? Or, how does this amount compare with the amount of CO_2 already in the atmosphere?

The warming of Earth's climate is a global phenomenon, and in analysing it, we must deal with quantities on a global scale, which means numbers that are much larger than those encountered in our daily lives. To gain some sense of scale, let us start with the size of a gigatonne. How much weight is a Giga tonne? *Giga* stands for a billion, that is, a thousand million. A Giga tonne is a thousand Mega (million) tonnes. Most of us are familiar with the notion of a tonne. A tonne is a unit of weight (mass) that stands for thousand kilograms[1] and a kilogram is the weight of one litre of water. A litre is a measure of volume. A thousand litres make up one cubic metre of water. So, a tonne is the weight of a cubic metre of water. Let us imagine a cubic metre of water as a 1m by 1m by 1m cube. What volume of water does it take to weigh a gigatonne? A kilotonne is the weight of a thousand one cubic meter blocks of water lined up to form a one-kilometre-long line. Packing a thousand such lines side by side gives us a lake of water one kilometre long, one kilometre wide and one metre deep, weighing one *Mega* tonne (1 million tonnes). Stacking another thousand such 1-metre-thick sheets of water, one on top of the other, gives us a cube of water one kilometre on each side. One cubic kilometre of water weighs one gigatonne! (1 billion tonnes).

[1]I will exclusively use the SI units throughout this book.

Mathbox 2.2.3

It is much easier to manipulate decimal numbers of vastly different magnitudes using a power-of-ten factor instead of writing long sequences of zeros on the left or the right of the decimal point. Writing out the number 37.04 Gt results in 37,040,000,000 tonnes. Instead we can write $37{,}040 \cdot 1000 \cdot 1000$ and abbreviate 1000 as $10 \cdot 10 \cdot 10 = 10^3$ resulting in $37040 \cdot 10^3 \cdot 10^3$; or even shorter as $37{,}040 \cdot 10^6$. Still shorter we write $37{,}040 \cdot 10^6 = 3.704 \cdot 10^4 \cdot 10^6 = 3.704 \cdot 10^{10}$. If we want to express the amount in kilograms, we have to multiply the number again by a thousand: $3.704 \cdot 10^{10} \cdot 10^3$ tonnes $= 3.704 \cdot 10^{13}$ kg, which is the weight in kilograms of 37.04 Gt written in *scientific notation.*

Now we can imagine that the weight of the 37.04 Gt of CO_2 emitted globally during 2019 amounts to the weight of a lake of about 6.1 km long by 6.1 km wide and a thousand metres deep! However, under normal conditions, carbon dioxide is a gas and, therefore, takes up much more space than liquid water. From school, we know that one mole of an ideal gas at standard conditions of temperature and pressure, 0° and atmospheric pressure, takes up 22.4 litres. One mole of a substance is a fixed, large number of molecules of that substance. The large number is *Avogadro's Number* equal to $6.0221408\ 10^{23}$. A mole of CO_2 weighs 44 g. If the volume of 44 g of CO_2 is 22.4 litres, what is the volume x of 37.04 Gt of CO_2? With a quick calculation, we get the answer:

If 44 g make 22.4 l, then 1 g makes $22.4/44 = 0.50909$ l, and 1kg makes $1000 \times 22.4/44 = 509.1$ l. The volume of the mass of CO_2 is over 500 times larger than that of the same mass of water. Therefore 37.04 Gt of CO_2 take up the volume

$$x = 3.704 \cdot 10^{13}\,\text{kg} \times \frac{22.4\,\text{l}}{0.044\,\text{kg}} \quad (2.1)$$

$$= 1885.7\ 10^{13}\,\text{l} \quad (2.2)$$

$$= 1885.7\ 10^{10}\,\text{m}^3 \quad (2.3)$$

$$= 18857\ \text{km}^3 \quad (2.4)$$

If we consider a one-kilometre thick layer, then the 18857 km^3 would cover an area of 18,857 km^2, which is equivalent to the area of a square of 137 km by 137 km, which is about half the area of Switzerland or slightly more than the total area of Kuwait. The amount of CO_2 emitted during 2019 would cover the whole area of Kuwait with a one-kilometre thick layer of CO_2.[2] Now that we know approximately how much space the emitted CO_2 takes up, we can ask the following question: how does the mass of the emitted CO_2 compare with the total mass of the atmosphere?

2.2.4 What Is the Mass of the Atmosphere?

We know the Earth's surface area. If we knew the mass of the air above a square metre of the Earth's surface, we could find the answer to our question by simply multiplying that mass by the Earth's surface in square metres. So, how can we know the mass of the air above one square metre of the Earth's surface? Well, mass is density times volume. We may take

[2]This is only approximately true because the pressure decreases with height and at 1 km altitude the atmospheric pressure would be lower than on the ground. Because CO_2 is a gas, it would expand and occupy more volume than the same gas mass on the ground. But this does not matter when all we want is to get a rough appreciation of how much volume the yearly emissions would take up.

the volume of a column of air up to, say, 30,000 m. The problem is that the air gets thinner and thinner at higher altitudes. The density of the air decreases with the height above sea level. That means, at higher altitudes, we have less air mass per unit volume. The density of air at one kilometre up is less than that at the surface, and ten kilometres up, it is even far less dense. Thus, it seems somewhat complicated to calculate the air mass over a square meter of ground area. In addition, over land, the Earth's surface is not at a uniform height due to hills and mountain ranges. Fortunately, there is a simpler way to estimate the atmosphere's mass because we know that the atmospheric pressure at sea level is, on average, 101,325 Pa (N/m^2).

The atmospheric pressure at sea level is the weight of the column of air above a unit area of the Earth's surface, which is the same as saying that the atmospheric pressure P at sea level is the weight F of the air mass M over an area A on the ground divided by the area A. The weight F is the mass M times the acceleration of gravity g. We have

$$P = \frac{F}{A} = \frac{M \cdot g}{A} \tag{2.5}$$

or

$$\frac{P}{g} = \frac{M}{A} \tag{2.6}$$

By dividing the pressure by the acceleration of gravity, we get the mass of air over one unit area, as follows:

$$\frac{P}{g} = \frac{101{,}325 \ \frac{\text{kg m}}{\text{s}^2\text{m}^2}}{9.81\frac{\text{m}}{\text{s}^2}} = 10{,}328\frac{\text{kg}}{\text{m}^2} = 10.328\frac{\text{t}}{\text{m}^2} \approx 1\ \text{kg/cm}^2 \tag{2.7}$$

Thus, the atmospheric pressure of 101,325 N/m^2at sea level corresponds to a mass of 10 tonnes of air above each square meter of the Earth's surface. From here, to get the atmosphere's mass, we only have to multiply that number by the Earth's surface. The surface area of a sphere is 4π times the square of its radius. The Earth's radius R is 6371 km, and thus the Earth's surface is

$$4\pi R^2 = 4 \cdot 3.14 \cdot 6371^2 \text{ km}^2 \tag{2.8}$$
$$= 12.566 \cdot 6371^2 \cdot 10^6 \text{ m}^2 \tag{2.9}$$
$$= 510.06 \cdot 10^{12} \text{m}^2 \tag{2.10}$$

And multiplying by the mass per square meter, we get that the mass of the atmosphere is:

$$510.06 \cdot 10^{12} \text{ m}^2 \cdot 10.328 \frac{\text{t}}{\text{m}^2} = 5.268 \cdot 10^{15} \text{ t} \tag{2.11}$$

Who would have thought that we could calculate the mass of the atmosphere on the back of an envelope?

2.2.4.1 An Afterthought

I calculated the mass of the atmosphere by using a simple model to make a prediction. First, I thought of the Earth as a perfect sphere, ignoring any surface irregularities such as mountains. Second, I assumed sea-level pressure to be the same everywhere, meaning the atmosphere is completely calm, with no wind, updrafts, or downdrafts that could affect surface pressure. Third, I used Newton's equations to relate surface pressure to the mass of air. How can I claim that my result is close to reality? Well, let us consider the effect of mountains. Mount Everest, the highest peak on Earth, rises to 8848.86 m above sea level. That is just below 9 km. The height of Mt. Everest is one and a half thousandths of the Earth's radius (8.85 km/ 6371 km = 0.0014). Thus, in approximating the Earth's surface as that of a perfect sphere, my error is less than 0.15%.

The lowest atmospheric pressure ever measured at sea level, in the centre of a tropical cyclone, is 870 hPa, and the highest,in Siberia, is 1084 hPa. The largest observed difference in atmospheric surface pressure is 1084 − 870 = 214 hPa, or 21% of the mean surface pressure. These extreme values occur sporadically, at different times, over small areas compared to the Earth's surface. The local difference between high pressure and low pressure is within 2%. But these variations are of little significance because the atmosphere's mass does not change with local pressure variations; therefore, we can safely use the long-term mean surface pressure to calculate the atmosphere's mass. Unfortunately, we do not have a direct method to measure the mass of the entire atmosphere to compare with our result. Therefore, we must rely on estimates, such as the one I provided above, and assess the potential errors associated with them. For my calculation, the errors are surely not larger than 1%, which will be sufficient for our purpose. The example also illustrates Occam's razor principle. Do not make your model more complicated than necessary.

2.2.5 Concentration of CO_2 in the Atmosphere

The estimated mass of CO_2 added to the atmosphere in 2019 was 37.04 Gt, which is a mass equal to

$$\frac{37.04\ 10^9\ \text{t}}{5.268\ 10^{15}\ \text{t}} = 7.03\ 10^{-6} \qquad (2.12)$$

of the previous mass of the atmosphere. If we mix the CO_2 with the air, we increase the per-weight concentration of CO_2 in the atmosphere by seven parts per million. Now, the number of 37.04 Gt of CO_2 added in 2019 means something.

The atmosphere is mainly nitrogen (around 80%), oxygen (around 20%), , and small amounts of CO_2 and other gases. We can also ask how much we increase the mass of CO_2 already present in the atmosphere. The *molecular* concentration of CO_2 in the atmosphere has been measured directly at the Mauna Loa Observatory in Hawaii since 1958.

In April 2019, the monthly average concentration of CO_2 in Earth's atmosphere was 413 parts per million in dry air (air without water vapour) in molar terms. That means out of a million molecules in the air, 413 are CO_2 molecules. The concentration we calculated above is on a weight (or mass) basis. Before we can compare our result with the molar concentrations measured at Mauna Loa, we must convert the weight fraction to a molar fraction.

To do the conversion, we recall that a mole of oxygen weighs 32 g ($2 \times 16 = 32$) and a mole of nitrogen weighs 28 g ($2 \times 14 = 28$), while a mole of CO_2 weighs 44 g ($12 + (2 \times 16)$). The weighted average molecular weight of the air mix of oxygen and nitrogen is

$$0.2 \cdot 32\,\mathrm{g} + 0.8 \cdot 28\,\mathrm{g} = 28.8\,\mathrm{g} \tag{2.13}$$

The molecular weight ratio of CO_2 to *air mix* is $44/28.8 = 1.528$. A given number of molecules of CO_2 weighs 1.528 times more than the same number of molecules from the air mix. If we now take away a fraction c of molecules from a mole of air mix and replace them with the same number of CO_2 molecules, then the total mass of one mole of the mixture is

$$m = (1 - c) \cdot 28.8\,\mathrm{g} + c \cdot 44\,\mathrm{g} = 28.8\,\mathrm{g} \\ - c \cdot 28.8\,\mathrm{g} + c \cdot 44\,\mathrm{g} = 28.8\,\mathrm{g} + c \cdot 15.2\,\mathrm{g} \quad (2.14)$$

The weight fraction of CO_2 in the mixture is

$$\frac{m_{CO_2}}{m} = \frac{c \cdot 44}{28.8 + c \cdot 15.2} \quad (2.15)$$

If the molecular fraction c is small, as in $c = 413\ 10^{-6}$, then contribution of the term containing c in the denominator is negligible

$$\frac{m_{CO_2}}{m} = \frac{c \cdot 44}{28.8 + 413\ 10^{-6} \cdot 15.2} \\ = \frac{c \cdot 44}{28.8 + 6.23\ 10^{-3}} \approx c \cdot \frac{44}{28.8} = 1.528 \cdot c \quad (2.16)$$

We can safely assume that the weight concentration of CO_2 in the air is 1.53 (1.528) times the molar concentration. We can now easily convert between molecular concentration and weight concentration. Our estimated weight concentration of 7.2 parts per million becomes $7.2/1.528 = 4.71$ parts per million molar concentration. For 2019, the *measured* increase in CO_2 in the atmosphere was 2.5 ppm, which is about half of what we calculated. Is our calculation wrong? Not at all. The reason is that, apart from the inherent uncertainties in all the data, the oceans absorb about half of the CO_2 released into the atmosphere by human activities. Our simple estimate allowed us to discover that there is more to the CO_2 cycle in the atmosphere than we initially thought. We discovered this by doing the numbers! The simple calculations capture many of the essentials, and you are now well prepared to make the most of the article "Carbon dioxide in Earth's atmosphere" in Wikipedia.

2.3 A Word About Energy and Its Units

The phenomenon of global warming is all about the sourcing and consumption of energy. The word "energy" is ubiquitous in our daily conversations, but for many of us, it remains somewhat mysterious. In physics, energy is a well-defined quantity that manifests itself in various forms. At the most fundamental level, mass equates to energy by Einstein's well-known equation $E = mc^2$, which tells us that mass can be transformed into energy, as happens in the explosion of a nuclear bomb. More importantly, the conversion of mass into energy is what powers the sun, which provides the energy for all life on Earth. Other than powering the sun, the conversion of mass into energy is not part of our daily experience. The forms of energy that we experience in our daily lives include thermal energy (heat), radiative energy (light), and kinetic energy (energy carried by a moving body, such as a passenger vehicle). In a collision, the kinetic energy of the moving vehicle deforms metal, breaks material, and produces heat. Here on Earth, by simply lifting an object from the ground up to a higher position, we can store energy. When we release the object, it falls to the ground, gaining kinetic energy that transforms into heat and deformation upon hitting the ground. Thus, in our minds, energy becomes this *ethereal fluid* that manifests itself in various ways with the fundamental property that in its transformation, its amount is always *conserved.* That means that if we have previously measured an amount of energy in one form, we will measure the same amount after it has been transformed into a new form.

The fact that energy is always conserved makes us think of energy as an *asset* that we use to do some *work* for us, such as warming our home or moving our car. We buy that asset (energy) in the form of fuel or in the form of *electricity* from *energy suppliers*. The fuel is the carrier of the *energy*. We can liberate the energy stored in the fuel by burning it. By *burning*, some of the energy stored in the chemical bond of the fuel substance is transformed into heat, which we may again transform into motion, or more generally speaking, into mechanical energy. In the case of electricity, we may transform the *electric energy* drawn from the power line into heat by a heating element or into mechanical work by an electric motor.

Energy is always conserved, but technical processes often waste some of it. For example, in an internal combustion engine, less than a third of the fuel's energy is converted to mechanical energy to move a car, while the rest is wasted as unusable heat. The useful energy from such a process is always less than what goes in. We characterise the conversion by its efficiency, which is the ratio of useful energy output to energy input.

2.3.1 Units of Energy

We want to know how much energy we can get from each source, who consumes it, and how much of it is consumed. There are several different units in use for measuring energy. I will adhere to the International System of Units, abbreviated as SI throughout this book. In the International System of Units, the unit of energy is the Joule. One Joule is quite a small amount of energy compared to our daily energy consumption. It takes 4180 J (4.18 kJ) to heat 1 litre of water by one degree. Starting at room temperature (20 °C), it takes nearly 568 kJ (over half a megajoule) to bring the water in a common-size (1.7 l) kettle to the boil. For comparison, a 1.8 m tall person weighing 80 kg does about 175 J of work in one push-up. Raising the temperature of one litre of water by one degree requires the energy equivalent of 24 push-ups.

Not surprisingly, expressing countries' and the world's yearly energy consumption in Joule yields large numbers. The world's annual energy consumption is approximately 600 EJ (Exa Joule). One Exa Joule is 10^{18} J. That is a one followed by eighteen zeros! The world's annual energy consumption in Joule is a number with 21 decimal digits. Your credit card number has 16 decimal digits. Energy consumption is often also expressed in terms of *power*. Power is the energy used per unit of time (second). The SI unit of power is the Watt. One Watt is one Joule per second. If your hot water kettle is rated 2000 W, it consumes 2000 J per second. In practice, it is easy to measure electric power consumption. Therefore, total energy consumption is often expressed in power multiplied by the time of its use, resulting in the familiar kilowatt-hour (kWh) unit. One kWh is the energy consumed when using 1000 W of electric power for one hour. Because we encounter energy consumption quoted in Watt (power), kWh, and Joule, we need to convert quickly from one unit to another. Table 2.1 gives the conversion factors.

Table 2.1 Energy and power conversions

Watt	× second	= Joule	= kWh
1	1	1	$0.2778 \cdot 10^{-6}$
1000	3600 (1 h)	$3.6 \cdot 10^{6}$	1
$31.701 \cdot 10^{9}$	$31.536 \cdot 10^{6}$(1 y)	$1 \cdot 10^{18}$ (1 EJ)	$0.2778 \cdot 10^{12}$
$1 \cdot 10^{9}$ (1 GW)	$31.536 \cdot 10^{6}$(1 y)	$31.536 \cdot 10^{15}$	$8.760 \cdot 10^{9}$

3

Life in the Greenhouse

In December 1968, William Anders took the now-famous photograph, reproduced in Fig. 3.1, of Earth from the Apollo 8 space capsule while orbiting the Moon. What differentiates this image from previous photos of the Earth taken from space is the Moon's barren surface in the foreground, which contrasts with a much more hospitable image of the Earth floating in the vastness of space. This image forever changed our perception of the world and begs the question: Why is the Earth so hospitable to us?

We might simply answer that it is because we are at the right distance from the sun, so that it is neither too hot nor too cold. This is true, but so it is for the moon; and, as far as we can see from Earth, the moon's surface is barren. In sunlight, the moon's surface temperature may reach 120 °C, hot enough to boil off any water present. But as soon as the sun sets, it gets bitterly cold down to −245 °C, where water and oxygen would freeze. So, there must be more to why the Earth's climate is so benign.

J. Sitte, *Global Warming: Facts, Myths and Remedies*,
https://doi.org/10.1007/978-3-032-20862-0_3

Fig. 3.1 *Earthrise* photograph taken on 24 December 1968 by William Anders from aboard the Apollo 8 spaceship orbiting the Moon (*source* NASA)

Clearly, the Sun's constant radiation is what warms the Earth. If the Earth absorbed and accumulated all the incoming solar energy, it would continue to warm indefinitely. But that is not what we observe. Instead, over millions of years, Earth's surface temperature has remained within a narrow range suitable for life. For this to happen, the Earth must get rid of the excess energy to prevent overheating. The balance between the incoming solar energy that warms the planet and the energy that the Earth returns to outer space determines the Earth's mean surface temperature. The mechanism that regulates this energy balance is known as the greenhouse effect. In this chapter, we will examine how the greenhouse effect operates and the crucial role greenhouse gases play in it.

3.1 Greenhouse Earth

Great thinkers of the 19th century, such as Fourier, Pouillet, Tyndall, Arrhenius,, and others, already pondered why the Earth has such a benign temperature. They all suspected it had to do with the CO_2 in the atmosphere. Here is what happens. Like all celestial bodies, the Earth and the Sun float in what we call outer, or interstellar, space. The Sun bathes the Earth in sunlight, warming it. How does the heat from the sun get here? It cannot be by thermal conduction. Interstellar space is almost entirely devoid of matter, and without matter, thermal energy cannot flow by conduction. However, sunlight is electromagnetic radiation, like radio waves, that travels unhindered through empty space and carries energy. The Earth absorbs some of the energy it receives from the Sun and reflects the rest back into space.

Feeding energy into a body heats it up. But, as far as we know, the mean temperature on the surface of the Earth was relatively stable over the last few millennia. That can only mean that Earth's thermal energy has not increased and that Earth returns all the radiated energy it absorbs from the Sun back into space. Because empty space does not conduct thermal energy, the only way Earth can return the excess energy to space is through radiation. Since we do not see the Earth glowing, we may not think it radiates energy. Yet, it does.

We know from physics that all bodies radiate energy, and that the rate at which they radiate increases with their temperature. With this knowledge, we can calculate the temperature of the Earth's surface at which the outgoing energy balances the incoming energy. According to this calculation, the Earth's mean surface temperature should be −18 °C. At −18 °C, water would be permanently frozen, and life could not have emerged. Instead, the mean surface temperature of the Earth is nearly 15 °C. So, why is the Earth warmer than our simple energy balance calculation predicts? What is wrong with our calculation of the equilibrium temperature?

The one thing ignored in the calculation is the layer of air that covers the Earth's surface. That layer of air, the atmosphere, must, therefore, act as a blanket that warms the Earth. Without that blanket, the Earth's surface would be exposed to extreme heat and cold, as the Moon's surface is.

Although early nineteenth-century thinkers could not yet calculate the energy balance, they already held the atmosphere responsible for keeping the Earth warm, much like a greenhouse captures solar radiation to warm its interior. Therefore, over time, the name *Greenhouse Effect* stuck for "how the atmosphere warms the Earth", although it differs somewhat from how the Sun warms the greenhouse.[1]

The Earth's surface and the lower atmosphere warm very much like the interior of a greenhouse. Although there is no physical roof, the atmosphere also functions like a glass roof, allowing sunlight in while preventing some of the thermal radiation from the Earth's surface from escaping back out. In addition, gravity holds the warm air back from drifting into outer space. In effect, the atmosphere is a blanket that creates the *climate* that allowed life to evolve on Earth.

3.2 Energy Balance in Space

Yet, even with the atmosphere, the Earth must return as much energy to space as it receives from the Sun to prevent its temperature from steadily rising. Reflection is one way to immediately return incoming radiation to space. The Earth reflects about 30% of incoming solar radiation. The remaining 70% is absorbed. Once solar electromagnetic radiation

---

[1] A greenhouse captures energy from the Sun by letting in the sunlight through a transparent glass or plastic roof. The solar radiation warms up the interior, but the glass roof and the walls do not let the warmth escape.

has been absorbed (captured) by the Earth, radiation is the only means to send some of it back to space.

Although a body radiates most of its energy around a peak-intensity wavelength, it also radiates at all other wavelengths. The intensity distribution across the emitted radiation's wavelengths closely follows an ideal distribution called the *black body* radiation spectrum. Interestingly, blackbody radiation depends only on the body's temperature, not on its material. Figure 3.2 shows the black body radiation spectrum for several temperatures. The spectra show that when the body's temperature decreases, the peak radiation intensity drops sharply. The peak also shifts towards longer wavelengths (lower frequencies).[2] For the Sun, which is very, very hot and glows in the sky with a temperature of nearly six thousand degrees Kelvin, the peak intensity of the radiation is in the *visible* range of the frequencies. This range is visible to us because our vision has evolved to take advantage of the abundance of solar radiation available in that frequency range.

For bodies at ambient temperature, the peak of their blackbody radiation spectrum lies in the infrared range , at much lower frequencies than the Sun's peak. The average temperature of the Earth's surface is 15 °C (288 K). Figure 3.3 compares the black body spectrum for 288 K with the black body spectrum for 6000 K. The peak of the 288 K body is at a wavelength of 10 μm, whereas the peak for the solar (6000 K) spectrum is at only 0.5 μm, and the intensity of the 288 K peak is much smaller. The Earth's peak radiation intensity is about half a million times weaker than the Sun's. The intensity of the Earth's 288 K spectrum had to be amplified by a factor of 500,000 to compare it with the Sun's 6000 K spectrum.

[2] The wavelength λ times the frequency ν, that is, the number of oscillations per unit time, is equal to the speed of light c. Thus $\nu = c/\lambda$.

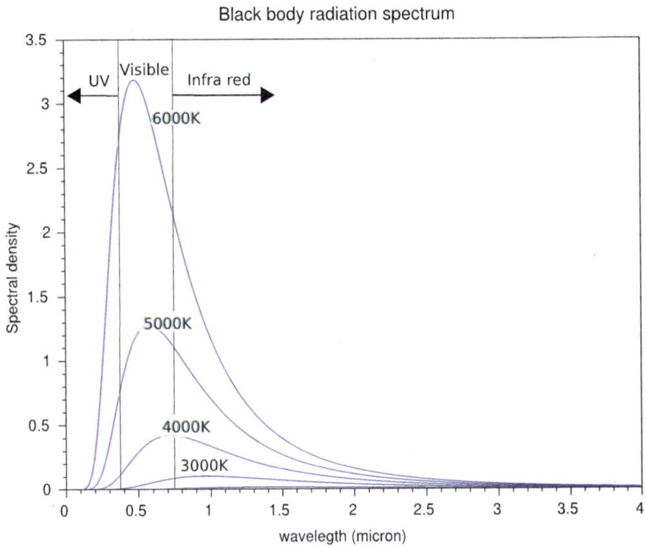

Fig. 3.2 Black body radiation spectra for high temperatures. The Sun's temperature is nearly 6000 K

We cannot see infrared light; therefore, we cannot see the infrared glow of the colder bodies. Although occasionally, if we get close to a warmer body, we may faintly feel the *glow* of its *thermal* radiation. We need specialized instruments, such as an infrared camera, to detect such bodies in the dark. In daylight, we see all these colder objects not by their thermal radiation but by the visible sunlight that they reflect towards our eyes.

All bodies lose energy through radiation and cool down unless the energy they lose is replenished. The filament in an incandescent light bulb cools down quickly once we switch off the electricity supply. The filament of an incandescent light reaches between 2000 and 3300 K, depending on its construction. From Fig. 3.2, we see that even at 3000 K, only a small part of the energy radiates as visible light; most of it

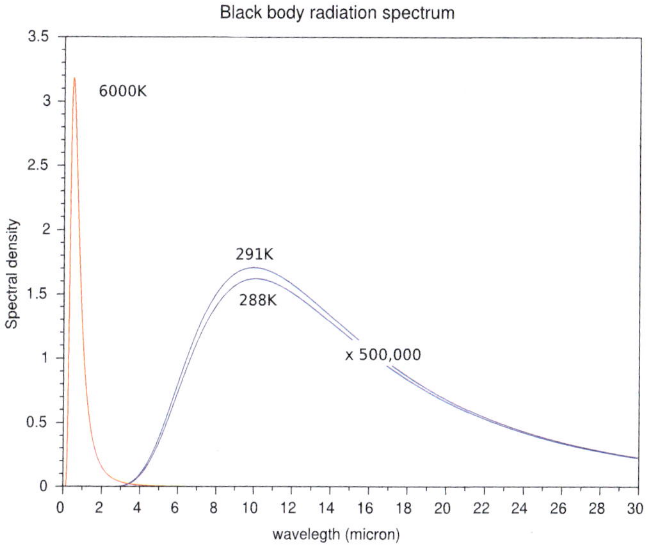

Fig. 3.3 Black body radiation spectra for Earth surface temperatures of 15° (288 K) and 18° (291 K)compared with the spectrum for 6000 K (sharp peak on the left). The Earth surface temperature spectra have been scaled up 500,000 times relative to the 6000 K spectrum

is in the infrared. Incandescent light bulbs have a vacuum inside to reduce heat loss from the filament by thermal conduction, and therefore, they only lose energy by radiation (well, mostly). Some other incandescent light bulbs contain an inert gas to reduce filament evaporation, thereby lengthening their lifespan. The price for this is a loss of energy by thermal conduction through the gas. In the Sun, the energy released by the inner nuclear fusion reactions replaces the energy lost through radiation. On Earth, incoming sunlight replenishes the energy radiated.

Because the blackbody radiation depends on the body temperature, the amount of energy the Earth radiates rises and falls with the Earth's mean surface temperature. The

Earth's surface cools if it radiates more energy than it receives from the Sun. The Earth's surface warms if it radiates less energy than it receives from the Sun. When the amount of energy radiated by the Earth balances the energy received from the Sun, the Earth's surface neither cools nor warms. Blackbody radiation maintains the Earth's surface temperature constant by causing it to radiate more energy as it warms, thereby counteracting the warming effect. Anything that disturbs the radiative energy flow from the Earth will shift the Earth's temperature, either higher or lower, until it reaches a new equilibrium.

3.3 Incoming Solar Energy

At the top of the atmosphere, every second, 1360 J of solar radiation energy pass through each square metre area facing the Sun. Consequently, the solar energy flux is $1.360\,\mathrm{kW/m^2}$.[3] Energy flux is the energy per second crossing a unit area perpendicular to the flow.

The radiative energy flux the Earth receives from the Sun is nearly constant, and this value is known as the solar constant.[4]

The value of the solar constant lets us easily calculate the total energy per second (power) that Earth intercepts from the Sun. We simply multiply the solar energy flux F by the area A of Earth's cross-section, as seen from the Sun.

[3]That is enough to keep a kettle going.

[4]The solar irradiation does change over time; it varies by about $1\,\mathrm{W/m^2}$ during the well-known 11-year sunspot cycle,, and it has undergone other oscillations over geological time scales.

The radius of the Earth is $R = 6378$ km; therefore, the cross-sectional area is

$$A = 2 \cdot \pi \cdot 6378^2 \,\text{km}^2 = 127.796 \; 10^6 \,\text{km}^2$$
$$= 127.796 \; 10^{12} \,\text{m}^2 \qquad (3.1)$$

Multiplying by the solar constant gives the total incoming power

$$127.796 \; 10^{12} \,\text{m}^2 \cdot 1.360 \,\frac{\text{kW}}{\text{m}^2} = 174.952 \; 10^{15} \,\text{kW}$$
$$= 0.175 \,\text{EW} \qquad (3.2)$$

Although the energy flux from the Sun is uniform over the Earth's cross-section, it is not uniform over the Earth's surface. The Earth is a sphere. Figure 3.4 shows that only the spot where the Sun is at the zenith (directly overhead) receives the complete solar flux. As we move away from this spot towards the periphery of Earth's face, the surface is no longer perpendicular to the Sun's radiation. One square metre of cross-section covers a larger area on Earth's surface. Area "B" in Fig. 3.4 shows this situation. The amount of energy crossing area "B" is the same as for area "A." However, it spreads over a larger surface area. As a result, the energy per unit area per second (flux) diminishes toward the rim of the illuminated face. At the rim, it is effectively zero. In short, the energy flux at a point on Earth's surface varies throughout the day, throughout the year, and with latitude.

At any moment, the Sun only illuminates half of Earth's surface. The Earth rotates at a uniform speed, so every point on its surface receives the same amount of exposure over time. Therefore, it is often convenient to use the average, over space and time, of the solar flux at the Earth's surface. To do this, we spread the flow that passes through Earth's cross-section over the entire surface. Since Earth's surface

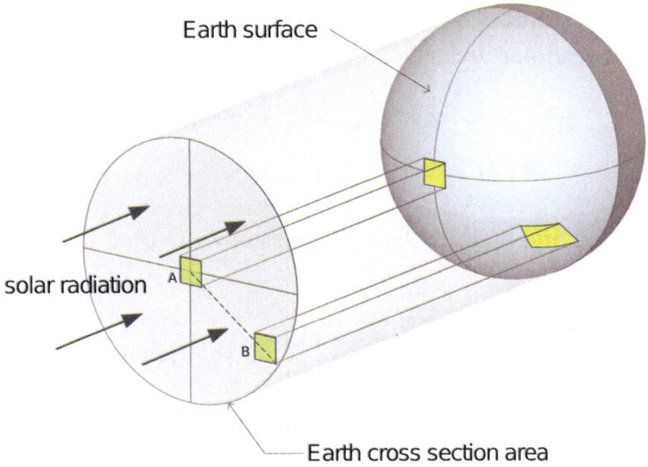

Fig. 3.4 Solar energy flux at the Earth's surface

area ($4\pi R^2$) is four times its cross-sectional area (πR^2), the solar flux intercepted by Earth spreads over four times the area. This reduces the average flux to one-fourth of the solar constant. Thus, the average radiative energy flux incident on the Earth above the atmosphere is $(1.360/4)\ \mathrm{kW/m^2} = 0.340\ \mathrm{kW/m^2}$. What happens to that flux next determines Earth's climate.

3.4 The Earth's Energy Balance

From the incoming sunlight, the Earth reflects almost one-third back into space and absorbs the rest. Both the atmosphere and the Earth's surface reflect and absorb sunlight. Thus, there are two parts to consider in the Earth's energy balance: the Earth's surface and the atmosphere. We see air as transparent and might think that all the sunlight reaches the ground. In reality, the Earth's atmosphere contains small

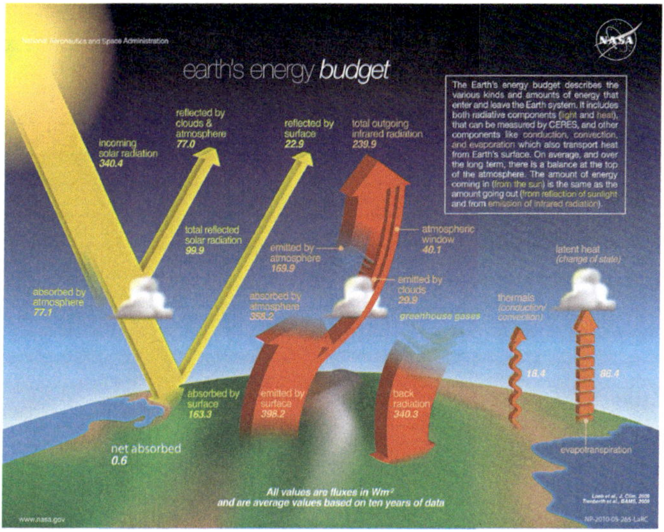

Fig. 3.5 Incident solar energy pathways

solid particles (dust) and liquid droplets (fog and clouds), making it only partially transparent. The atmosphere absorbs or reflects nearly half of the incoming solar radiation. The other half reaches the Earth's surface, where it is reflected and absorbed. To understand what happens to all the solar energy, we need to follow each of the four pathways.

NASA's Earth Observatory (Lindsey 2012) has measured what happens to the incoming solar energy flux and visualised the results in the poster shown in Fig. 3.5. The incoming flux is distributed across the four pathways of absorption and reflection by the atmosphere and by the ground, listed in Table 3.1,

The data show that nearly one-third (29%) of the incoming radiation is reflected into space. The atmosphere reflects almost three times as much as the Earth's surface. Adding the amount of solar radiation absorbed by the atmosphere to the amount reflected by the atmosphere reveals that only

Table 3.1 Earth's incoming energy budget

Incoming energy pathway	Flux W/m^2	Percentage
Reflected by clouds and dust	77.0	22.6
Reflected by the surface	22.9	6.7
Total reflected	99.9	29.3
Absorbed by Earth's atmosphere	77.1	22.6
Absorbed by earth' surface	163.4	48.0
Total absorbed	240.5	70.7
Total	340.4	100

Data from NASA Earth observatory

Table 3.2 Earth outgoing radiation energy budget

Outgoing energy pathway	Flux W/m^2	Percentage
Radiated into space directly from the surface	40.1	11.8
Radiated by the clouds	29.9	8.8
Radiated by earth's atmosphere	169.9	49.9
Total	239.9	70.5

Data from NASA Earth observatory

about half of the incident flux reaches the Earth's surface. The Earth's surface absorbs most of the radiation that reaches it and only reflects a small part. Most of the absorbed incoming energy replaces the heat lost by the atmosphere and the ground through infrared radiation. A small portion of the adsorbed solar energy (0.14%) is directly converted and stored as chemical energy through photosynthesis in plants.

Table 3.2 shows that only a small part of the energy absorbed by the ground radiates directly back into outer space. That is the part of the radiation from the Earth's surface that can reach outer space by passing through the *atmospheric window* transparent to infrared radiation. The atmosphere and clouds contribute most to maintaining the energy balance by radiating energy back into space.

Table 3.3 Energy balance of the Earth's atmosphere

Energy source/sink	Flux W/m^2
Radiation from surface absorbed by atmosphere	358.2
Absorbed from sunlight	77.1
Evaporation of water	86.4
Thermal conduction	18.4
Radiated back to ground	−340.3
Radiated by the clouds	−29.9
Radiated into space	=169.9
Total	0.0

Data from NASA Earth observatory

Most of the energy radiated by the Earth's surface heats the lower part of the atmosphere. The next-largest energy input to the atmosphere is the energy stored in water evaporated (clouds) from the surface and emitted by vegetation and animal transpiration. The latter is slightly larger than the energy the atmosphere absorbs directly from incoming sunlight. Finally, the atmosphere receives a small amount of energy from the Earth's surface through direct thermal conduction. Table 3.3 gives the measured values for these various energy inputs to the atmosphere.

The atmosphere dissipates all the energy it receives by radiating nearly two-thirds back to Earth and returning the remaining third into space. The upper part of the atmosphere radiates energy directly into space. The heat in the lower atmosphere drives the climate. The atmosphere acts as a blanket to the Earth, regulating the balance of incoming and outgoing energy flux. The atmosphere makes Earth's biosphere a greenhouse.

Table 3.4 shows that the large amount of energy the atmosphere radiates back to the Earth dominates its energy budget. Any change in that amount will alter the balance and, therefore, the surface temperature. Thus, we must ask what

Table 3.4 Energy balance for the Earth (surface)

Energy source/sink	Flux W/m^2
Sunlight absorbed by the Earth's surface	163.4
Radiated to ground from the atmosphere	340.3
Radiated by earth's surface	−398.2
Evaporation of water	−86.4
Thermal conduction	−18.4
Total	0.7

Data from NASA Earth observatory. The small surplus is captured as chemical energy by plants and animals

can cause a change in the energy radiated back to the surface by the atmosphere.

3.5 Greenhouse Gases

The atmosphere consists mainly of nitrogen (78%) and oxygen (21%). The rest consists of argon (1%), water vapour, and traces of carbon dioxide, methane, ozone, and other gases. Nitrogen, oxygen, and argon, the three gases that comprise nearly all of the atmosphere, are transparent to infrared radiation. However, there are other gases present in small concentrations that absorb electromagnetic radiation at wavelengths where the Earth's infrared radiation index-infrared radiation is strongest. The infrared absorption by these gases reduces the infrared radiation from the Earth that can escape directly into outer space, thereby reducing the amount of incoming solar energy that is returned into outer space. The gases that contribute most are water vapour, carbon dioxide, ozone, and methane. These *greenhouse gases* capture some of the energy radiated by the Earth's surface and distribute it into the atmosphere, which will then radiate it back to the surface. An increase in the amount of greenhouse gases in the atmosphere will increase the energy radiated back to the surface, thereby raising the temperature at the Earth's surface and in the lower atmosphere. As the Earth's surface temperature rises, so does its thermal radiation, until it reaches a new equilibrium at a higher temperature. **Thus, the amount of greenhouse gases in the atmosphere sets the Earth's surface temperature in the same way as a thermostat sets a room heater's heat output.**

3.5.1 Infrared Absorption Spectra of Greenhouse Gases

The infrared absorption spectrum of a gas describes its infrared absorption properties in detail. To obtain an infrared absorption spectrum, one needs an infrared spectrometer that shines a beam of infrared light through a gas sample (in a container transparent to infrared radiation) and measures how much radiation the sample absorbs at each wavelength. Figures 3.7, 3.8, 3.9 and 3.10 show the infrared absorption spectra of water vapour, CO_2 methane and ozone. The blue trace in these figures displays infrared spectra as spectroscopists usually do. They plot the fraction of infrared light let through by the sample against increasing wavelength (decreasing frequency), as in blackbody radiation spectra plots, but expressed in wave numbers. The fraction of infrared light let through by the sample is the *relative transmittance.* The wave number is the number of wavelengths that fit in a centimetre. Decreasing wave number means increasing wavelength and thereby decreasing frequency.[5] Wave numbers take a convenient range of numbers, from

[5]The frequency ν of an electromagnetic wave is the speed of light c divided by the wavelength λ : $\nu = \frac{c}{\lambda}$ or $\frac{\nu}{c} = \frac{1}{\lambda}$. By dividing the frequency by c and expressing the wavelength in centimetres, we get the wave number ($\frac{1}{\lambda}$), which is proportional to the frequency.

4000 cm^{-1} (wavelengths per centimetre) to 500 cm^{-1}, for the frequency range of interest.

The infrared absorption spectra of gases exhibit clusters of many narrow, inverted peaks connected by regions with minimal absorption. The inverted peaks are absorption *lines* that cluster together into absorption *bands*. Each line corresponds to a frequency at which gas molecules can absorb infrared radiation by jumping to a higher energy quantum state of vibration and rotation. The area under the spectral curve is proportional to the energy let through by the sample over the selected spectral range. The area above the curve, between it and the relative transmittance = 1.0 line, is proportional to the energy absorbed by the sample.

To see how the greenhouse gases trap outgoing radiation, we first plot, in Fig. 3.6, the Earth's blackbody radiation of Fig. 3.3 as a function of wave numbers, instead of wavelength, over the range where the greenhouse gases absorb infrared radiation.

Once we find the range of wave numbers at which the Earth emits its blackbody radiation, we can combine the Earth's emission spectrum (black trace) with the infrared absorption spectrum (blue trace) of a greenhouse gas. This shows where the absorption reduces the radiation passing through the atmosphere. The Earth's blackbody radiation curve (black trace) indicates the amount of infrared radiation emitted by the Earth into the atmosphere. The infrared absorption spectrum (blue trace) shows the fraction of incident radiation absorbed by the corresponding gas. By multiplying the Earth's blackbody emission spectrum by the infrared absorption spectrum of each greenhouse gas, we see the amount that is let through (red trace) at each frequency by the greenhouse gas present in the atmosphere. The amount of energy absorbed by each gas depends on the gas's concentration, its ability to absorb infrared radiation, and the length of the path the radiation travels through

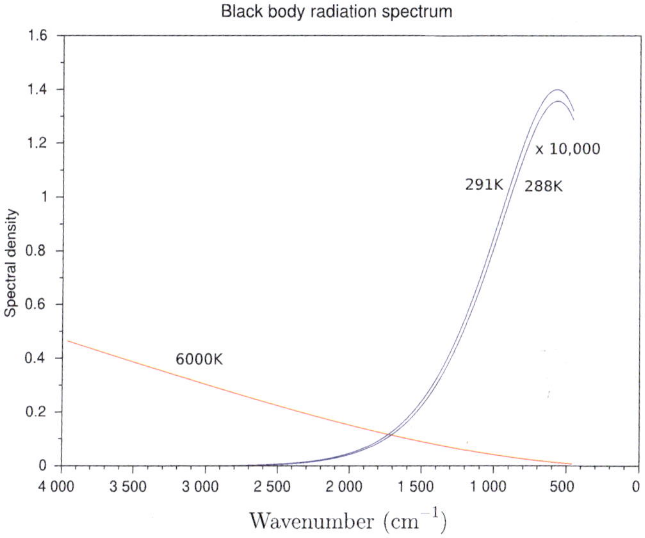

Fig. 3.6 Black body radiation spectra of the Earth's surface from Fig. 3.3 plotted as a function of wave number over the range where greenhouse gases absorb infrared radiation

the atmosphere. I do not take these factors into account in the graphs to keep it simple and focus on illustrating how absorption by greenhouse gases reduces the Earth's infrared radiation that reaches outer space.

With that understanding, we will now examine the roles of the main greenhouse gases.

3.5.2 Water Vapour

Water vapour is, after Oxygen and Nitrogen (and Argon), the most abundant gas in the atmosphere. However, water vapour content varies widely over space and time. Think of clouds and local humidity. In the tropics, humidity can be high, with up to 5.

What differentiates water vapour from other greenhouse gases is that it can condense in the atmosphere and fall out of the atmosphere as rain. CO_2 and Methane do not condense (liquefy) in the range of pressures and temperatures found in the atmosphere. At constant pressure, the air's temperature determines how much water vapour it can hold. The maximum amount of water vapour the air can hold (100% humidity) is the saturation point. The saturation point rises with the temperature. When humidity is high and the temperature drops, the saturation point falls, and excess water condenses. Most of the time, the local water vapour content of the atmosphere is below saturation.

The infrared spectrum of water in Fig. 3.7 shows that water is a strong absorber of infrared radiation. As a strong infrared absorber and the greenhouse gas with the highest atmospheric concentration, water vapour is the most potent of the greenhouse gases. It plays a vital role in the local weather, but because it can condense out of the atmosphere and does not accumulate over time, it only plays a secondary role in the long-term warming of the atmosphere. Still, water vapour provides a positive feedback mechanism for atmospheric warming, in that as the temperature rises, there is more water vapour in the atmosphere, leading to increased infrared absorption and further warming. On the other hand, more clouds mean more reflection of incoming solar radiation, which reduces the temperature.

3.5.3 Carbon Dioxide

The infrared spectrum of CO_2 in Fig. 3.8 shows two strong absorption bands. One band is at a wavelength where the Earth's infrared radiation is very low. The other absorption band occurs at a wavelength where Earth's infrared radiation is near its maximum, meaning there is abundant infrared

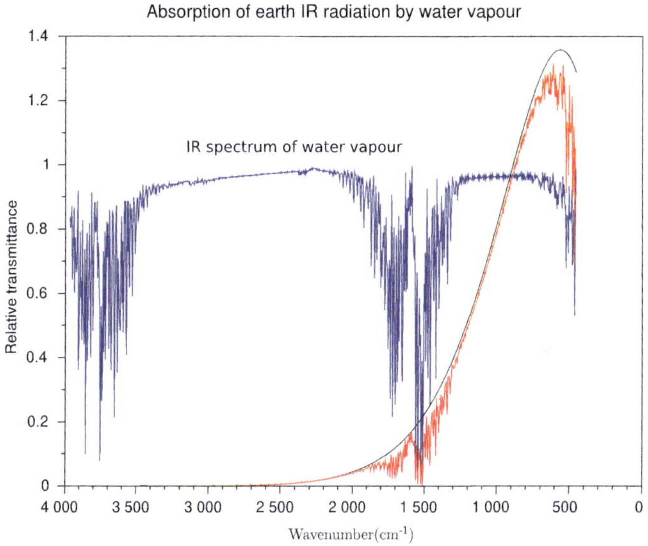

Fig. 3.7 Infrared relative absorption spectrum of water vapour. For comparison, the continuous curve on the right is proportional to the Earth's blackbody radiation spectrum at 288 K. The spectrum underneath is the Earth's blackbody radiation spectrum multiplied by the water vapour IR spectrum, to illustrate how absorption by water vapour reduces the Earth's blackbody radiation as it passes through the atmosphere

energy for CO_2 to absorb, and that CO_2 is an important greenhouse gas.

By the end of 2023, the CO_2 molecular concentration in the atmosphere was 422 parts per million (ppm) or 0.04%, which is ten times less than the average atmospheric water vapour concentration. However, carbon dioxide is removed very slowly from the atmosphere; it lingers for a long time, indeed for centuries. Therefore, the CO_2 emitted by the combustion of fossil fuels accumulates in the atmosphere. That makes CO_2 the largest contributor to long-term atmospheric warming by far.

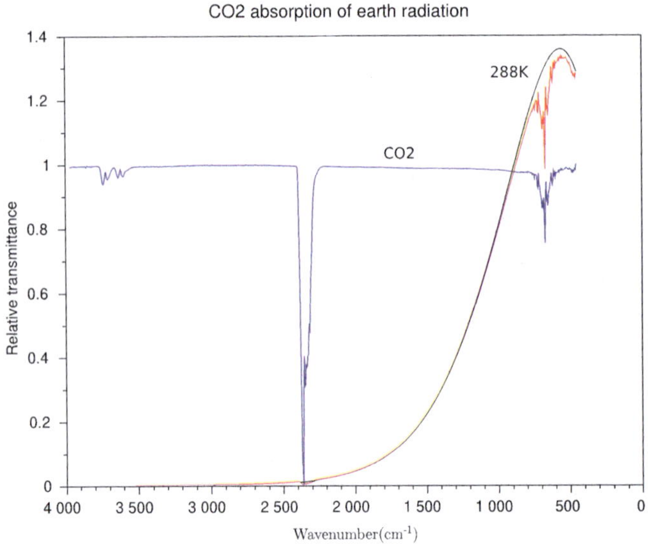

Fig. 3.8 Infrared relative absorption spectrum of CO_2. For comparison, the continuous curve on the right is proportional to the Earth's blackbody radiation spectrum at 288 K. The spectrum underneath is the Earth's blackbody radiation spectrum multiplied by the CO_2 IR spectrum, to illustrate how absorption by water vapour reduces the Earth's blackbody radiation as it passes through the atmosphere

3.5.4 Methane

Methane is the next most abundant greenhouse gas in the atmosphere after water vapour and carbon dioxide. Methane does not linger in the atmosphere as long as CO_2. Natural processes remove Methane from the atmosphere, giving it a lifetime of about 12 years in the atmosphere. With a molecular concentration in the Earth's atmosphere close to 2000 parts per billion (1902 ppb at the end of 2023), the molecular concentration of Methane is about half a per cent of the 422 parts per million (ppm) molecular concentration of CO_2.

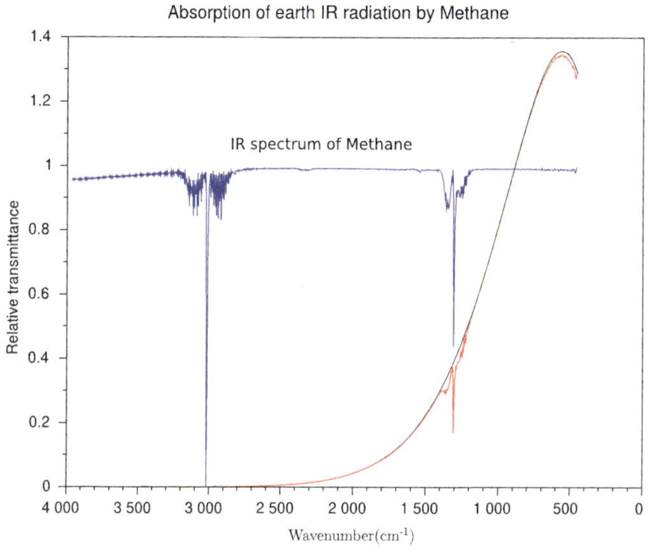

Fig. 3.9 Infrared absorption spectrum of methane (CH_4). For comparison, the continuous curve on the right is proportional to the Earth's blackbody radiation spectrum at 288 K. The spectrum underneath is the Earth's blackbody radiation spectrum multiplied by the methane IR spectrum, to illustrate how absorption by water vapour reduces the Earth's blackbody radiation as it passes through the atmosphere

Methane is the simplest of the hydrocarbon compounds. A Methane molecule consists of four hydrogen atoms bound to one carbon atom, giving it the chemical formula CH_4.

The infrared spectrum of Methane in Fig. 3.9 shows that, like CO_2, it also has two absorption bands in the infrared. The absorption band of Methane that falls within the spectral range of Earth's infrared radiation occurs at a wavelength where Earth's infrared radiation is approximately one-third of its peak value. The energy available for the Methane to absorb is only one-third of the energy available for CO_2 to absorb.

Methane's contribution to atmospheric warming is greater than its low concentration would suggest because, per molecule, its capacity to absorb infrared radiation is about 50 times greater than that of CO_2. Detailed calculations of the net contribution of Methane to warming in comparison with CO_2 are complicated by its higher absorption, the position of its absorption bands relative to the Earth's infrared emission spectrum, its overlap with water vapour absorption bands, its much lower concentration, its shorter lifetime in the atmosphere, and its reactivity with other atmospheric gases. In Sect. 6.6.1, we examine the methane contribution to warming more closely because the fossil fuel industry uses the higher infrared absorption of Methane compared to CO_2 to divert attention from the contribution by CO_2, and environmentalist groups use it to campaign against the livestock industry.

The sources of atmospheric Methane are different from those of CO_2, and the history of the increase of the methane concentration in the atmosphere is also different, as is the time it lingers in the atmosphere. Therefore, the reduction of Methane in the atmosphere is a problem distinct from that of the CO_2 reduction (Ravishankara et al. 2021; Kühne et al. 2022). However, the warming of the atmosphere accelerates the emission of Methane from natural sources and reduces the atmosphere's capacity to remove methane (Cheng and Redfern 2022). The main single contributor to anthropogenic methane emissions is the fossil fuel industry, accounting for 33%, compared with 27% from livestock. Methane emissions from the fossil fuel industry come from Methane released during the mining, transportation, and processing of fossil fuels. The remaining 40% come from landfills, rice cultivation, and burning of biomass and biofuels.

3.5.5 Ozone

Ultraviolet (UV) radiation from the sun produces ozone (O_3)—an unstable oxygen molecule—in the upper atmosphere. The ozone infrared spectrum in Fig. 3.10 displays an absorption band near 1000 cm^{-1}, located on the high-frequency shoulder of the Earth's black body radiation spectrum. It isn't easy to obtain an IR spectrum of pure ozone. Therefore, the spectrum in Fig. 3.10 also contains absorption lines from contaminants in the sample. The most notorious are the CO_2 bands around 2350 and 675 cm^{-1}. Ozone is essential to life on Earth. Ozone reduces the damaging UV radiation from the sun that reaches Earth's surface. Ozone concentration in the upper atmosphere is nearly constant. Therefore, it plays no role in long-term changes in the radiation balance.

3.5.6 Earth's Radiation Absorbed by the Atmosphere

To better understand how greenhouse gases modify the blackbody radiation emission spectrum of the Earth, that is, how it appears when viewed from above the atmosphere, we need to include the effects of the main greenhouse gases. I have done this for the black trace in Fig. 3.11. Here, each gas has been given equal weight by multiplying the blackbody spectrum by the product of the infrared absorption spectra of water vapour, CO_2, and Methane, disregarding the individual concentrations and absorption propensities. To compare the graph with the Earth's infrared radiation emission flux measured by the Nimbus 4 satellite above the atmosphere (black line), shown in Fig. 3.12, I plot the curves against *increasing* wave number (the reverse of as in the figures above).

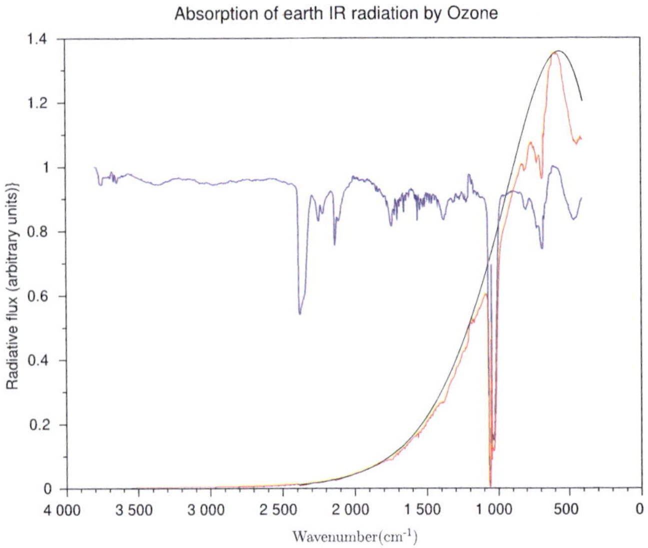

Fig. 3.10 Infrared absorption spectrum of O_3 (ozone).For comparison, the continuous curve on the right is proportional to the Earth's blackbody radiation spectrum at 288 K. The spectrum underneath is the Earth's blackbody radiation spectrum multiplied by the O_3 IR spectrum, to illustrate how absorption by water vapour reduces the Earth's blackbody radiation as it passes through the atmosphere

Our predicted Earth infrared radiation spectrum (Fig. 3.11), measured above the atmosphere, qualitatively agrees with the spectrum measured by the Nimbus 4 satellite (Fig. 3.12).

The spectrum measured by Nimbus 4 shows a strong CO_2 absorption band near the peak of the Earth's radiation. Broad absorption *shoulders* from water vapour appear at both sides of the peak. On the high-frequency shoulder, we observe a region where there is almost no difference between the blackbody spectrum and Earth's infrared spectrum, indicating that Earth's radiation passes through the atmosphere without being absorbed. This region is the

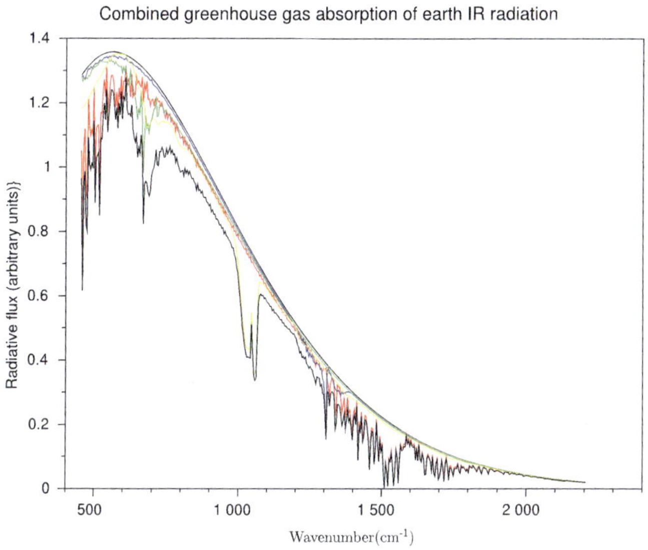

Fig. 3.11 Qualitative combined Earth IR radiation absorption by water vapour, carbon dioxide, and Methane

atmospheric window where the atmosphere is transparent to infrared radiation. The window is only interrupted by an ozone (O_3) absorption band. Further along the shoulder towards higher frequency, the methane absorption band overlaps with those of water. At that position in the spectrum, Earth's radiation has already decayed to about 40% of its peak intensity, and water vapour's absorption further reduces it. The lower Earth radiation intensity at the methane absorption band compensates somewhat for Methane's much higher infrared absorption per molecule than carbon dioxide.

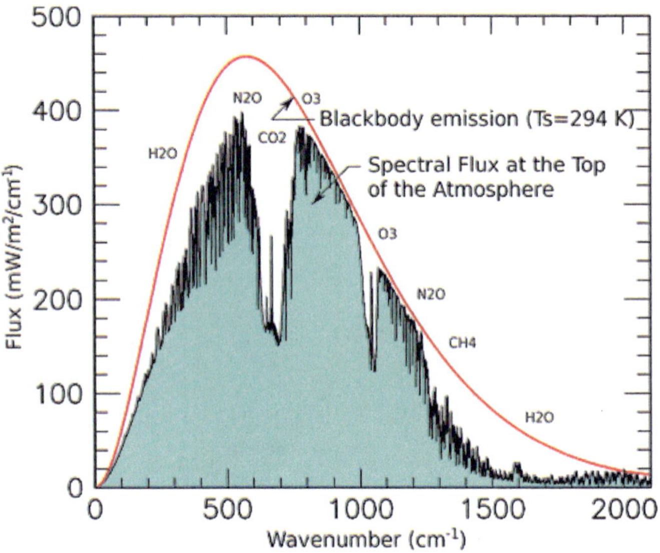

Fig. 3.12 Example of the spectrum of infrared radiation flux from the Earth measured above the atmosphere by NASA Nimbus 4 satellite (Hansel 1994). The smooth line is the ideal spectrum of the radiation emitted by a body at 294 K (21 °C)

3.6 Radiative Forcing

The spectrum, shown in Fig. 3.12, of the outgoing radiation flux (Watt per m^2) measured above the atmosphere illustrates how greenhouse gases absorb outgoing radiation, reducing the Earth's outgoing radiative flux. *Radiative forcing* is the name given to the *change* in radiative flux in the atmosphere due to a perturbing factor, such as a change in greenhouse gas concentration. Thus, radiative forcing is the crucial quantity for estimating the change in the Earth's surface temperature resulting from a disturbance.

To think in quantitative terms, let us define the net flux F_n as the difference between the downward (incoming) flux F_i and the upward (outgoing) flux F_o.

$$F_n = F_i - F_o \tag{3.3}$$

A change in the incoming flux, a change in the outgoing flux, or both changes the net flux. An increase (decrease) in solar radiation or a decrease (increase) in the concentration of atmospheric gases that absorb solar radiation increases (decreases) the incoming flux F_i. An increase (decrease) in the concentration of greenhouse gases decreases (increases) the outward radiation flux from the Earth's surface F_o.

The change ΔF_n in net flux due to a perturbing factor

$$\Delta F_n = \Delta F_i - \Delta F_o \tag{3.4}$$

is the radiative forcing. A positive value of ΔF_n, that is, positive radiative forcing, can come about when the incoming flux is unchanged ($\Delta F_o = 0$), and an increase in greenhouse gas concentration reduces the outgoing flux ($\Delta F_o < 0$)

Measurements by the Nimbus 4 satellite of the spectral composition of the outgoing radiation flux at the top of the atmosphere help separate the contributions of each greenhouse gas to the reduction in outgoing flux. However, these measurements alone do not suffice to determine the radiative forcing, and we need to rely on calculations based on laboratory measurements. Such calculations are complicated because one must add the amount of infrared radiation absorbed by the many absorption bands from each gas. The absorption by each gas depends on its concentration and the effectiveness of its molecules in absorbing the radiation they receive (the absorption cross-section).

NOAA publishes estimates for radiative forcing in the Annual Greenhouse Gas Index (AGGI) (NOAA 2021). The NOAA values for radiative forcing are the changes in flux since the start of the Industrial Era, as defined by the IPCC as 1750. For 2020, the radiative forcing from the change of

CO_2 content in the atmosphere was 2.111 W/m^2, and for Methane it was 0.520 W/m^2. According to NOAA, between 1990 and 2022, the total radiative forcing increased by 49%. CO_2 contributed 78% to this increase. Methane and nitrous oxide N_2O contributed nearly equally, with about 7.5% each.

4

Global Warming

In the previous chapter, our analysis of the greenhouse effect and Earth's energy balance demonstrated that atmospheric greenhouse gases play a crucial role in regulating the Earth's average surface temperature. Among these gases, CO_2 leads in shifting the balance between the incoming radiation flux from the sun and the outgoing radiation flux from the earth.

Since the start of the industrial era, in the mid-18th century, humans have burned ever-increasing amounts of fossil fuels, and thereby dumped ever more CO_2 into the atmosphere. Part of this CO_2 has stayed in the atmosphere, shifting the radiation balance towards higher temperatures. How big is the temperature shift, and more importantly, what is the future temperature rise that we can expect from continuing with the dumping of CO_2 into the atmosphere? In this chapter, we will work out an answer to this question.

J. Sitte, *Global Warming: Facts, Myths and Remedies*,
https://doi.org/10.1007/978-3-032-20862-0_4

4.1 Our Current Predicament

The amount of warming produced by humankind's CO_2 emissions has been the subject of intense research by climate scientists in recent decades. The 6th Assessment Report by the IPCC (Masson-Delmotte, 2021) summarises the results of this research. According to the IPCC 6th Assessment Report

> Human activities, principally through emissions of greenhouse gases, have unequivocally caused global warming, with global surface temperature reaching 1.1 °C above the 1850–1900 mean in 2011–2020.

The historical data show that the accumulated CO_2 emissions since 1850, estimated at 2300 Gt, rose the average surface temperature by 1 °C, giving an increase of 0.27–0.63 °C per 1000 Gt CO_2 with a best estimate of 0.45 °C. (Masson-Delmotte, 2021)

Climate scientists employ sophisticated mathematical climate models of the Earth's climate to predict future temperature increases and apply these models to various scenarios of future fossil fuel consumption, ranging from optimistic to pessimistic. These mathematical models are bottom-up models that try to simulate the multitude of processes and their interactions that give rise to the climate. Consequently, these models get rather complicated and are almost impossible for laypeople, and even for many experts, to grasp. Fortunately, everyone can make sound forecasts with a few simple calculations by examining how atmospheric CO_2 concentration has grown over time and comparing it with the global mean surface temperature record. Let us start by inspecting the data.

4.2 Historical Global Temperature Data

Reliable weather prediction greatly benefits our economic activities and personal lives. Measuring and recording the ambient temperature is the first step toward establishing a quantitative foundation for weather prediction. Regular, worldwide temperature measurements are an essential quantitative foundation for both short-term weather forecasts and climate change predictions.

Reproducible temperature measurements became possible in the early 18th century when Daniel Gabriel Fahrenheit and others perfected the thermometer. However, widespread thermometer recordings began only in the mid-19th century. This coincided with the rising use of fossil fuels during the Industrial Revolution and was spurred by concerns about human-induced climate change, which was then feared to result from deforestation. (There is the exceptional Central England Temperature (CET) record, which gives monthly mean temperature records since 1659.) This shift to systematic measurement marks a pivotal point in the development of our modern understanding of climate.

It is also possible to estimate surface temperatures in earlier times, including prehistoric and geological periods, using indirect methods such as analysis of ice cores , tree rings, and sediment layers. These estimates show that the Earth's climate fluctuated significantly over time. However, it is sufficient to examine only the recent past to assess the impact of CO_2 emissions on the Earth's climate. Thermometer measurements collected over nearly two centuries show that the planet is warming.

4.2.1 The Global Mean Surface Temperature

The weather stations that take temperature measurements are mainly on the ground. Therefore, our records show the temperature close to the Earth's surface. The *surface temperature* varies widely during the day, the seasons, and across geographic locations. Surface temperatures on land are generally higher than at sea. Temperatures in towns and cities are higher than in the countryside. Additionally, the higher in the atmosphere, the greater the temperature difference from the surface. Even so, there is general agreement that the Global Mean Surface Temperature (GMST) provides a helpful characterisation of the Earth's climate. Surface temperature is the variable that is easiest to measure continuously and simultaneously worldwide. Last but not least, the surface temperature is what we feel.

Measurements from over 20,000 weather stations on land, ships, and buoys provide the data used to calculate the Global Mean Surface Temperature. In addition to measuring temperature with thermometers, we now have several remote sensing techniques for measuring surface temperature using satellites that contribute to the data.

There are several methods to compute a GMST from all this data. Accordingly, several data sets are available for the time series of global mean surface temperatures. Surprisingly and reassuringly, their values coincide very closely. Here, I will use the NOAA data set. The NOAA data set lists the monthly global temperature *anomaly* , defined as the global mean surface temperature deviation from the 100-year average from 1901 to 2000. The 100-year average global mean surface temperature from 1901 to 2000 is 13.9 °C.

The graph, in Fig. 1.1, of the NOAA year-by-year GMST anomaly for October, since 1850, with the 100-year mean added, shows that despite year-to-year fluctuations, the

GMST is slightly on the rise.[1] The graph in Fig. 4.1 of the same data but without the 100-year average added, shows the temperature rise more clearly.

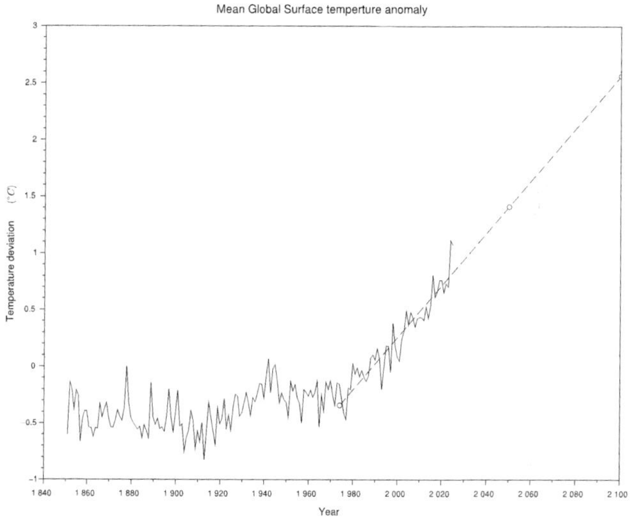

Fig. 4.1 Year by year deviation of the global mean surface temperature from the 100-year average from 1901 to 2000 for the month of October (solid line) in the NOAA data set. The dashed line is a linear extrapolation to the end of the 21st century of the temperature data between 1974 and 2024

For 2024, the October temperature increase above the 100-year average (1901–2000) was 1.071 °C. From around 1974 onwards, the rate of temperature rise was steeper than earlier. For this period, the curve seems to follow a linear trend. One can calculate the best straight-line fit (least-squares) to the data points from 1974 to 2024. My result is the dashed line in Fig. 4.1. This result gives us an

[1]The fluctuations in mean surface temperature are not only due to measurement uncertainty but to various smaller and shorter duration processes such as volcanic eruptions, variations in solar output or sulphate aerosol pollution.

average yearly increment to the mean surface temperature of 0.02302 °C. If this trend persists, we can predict by linear extrapolation that by 2050, the temperature rise will reach 1.408 °C above the 20th-century average, and that by the end of the century, it may rise by 2.56 °C.

4.2.1.1 The Temperature in the Pre-Industrial Era

The Paris Accord set the goal of limiting the global temperature rise to 1.5 °C above *pre-industrial* levels. But what is the pre-industrial level? The pre-industrial era ended with the widespread use of machinery for industrial production, which historians agree began around the middle of the 18th century (1750). As mentioned above, systematic temperature recording began only a hundred years later, around 1850. Due to the lack of detailed temperature records prior to 1850, the IPCC used the 1850–1900 average of observed global surface temperatures as the pre-industrial level. This average is 0.27 °C below the 20th-century average used in the NOAA data set as the reference level for the temperature anomaly. When we reference the NOAA data to the accepted pre-industrial temperature average, we must add 0.27. With this addition, the increase for October 2024 becomes 1.34 °C and the extrapolated future values rise to 1.68 °C for 2050 and 2.83 °C for the end of the 21st century.

4.3 Is CO_2 the Causes of the Temperature Rise?

Our current understanding of the greenhouse effect is well grounded in the laws of physics. It predicts that greenhouse gases regulate the radiation balance and, hence, the global mean surface temperature of the Earth. On this basis, I have taken for granted that the cause of the observed rise in GMST is the massive emission of CO_2 from human activity since the Industrial Revolution. Indeed, an overlay of the Keeling curve and the global mean surface temperature record in Fig. 4.2 shows that the two curves move in tandem.

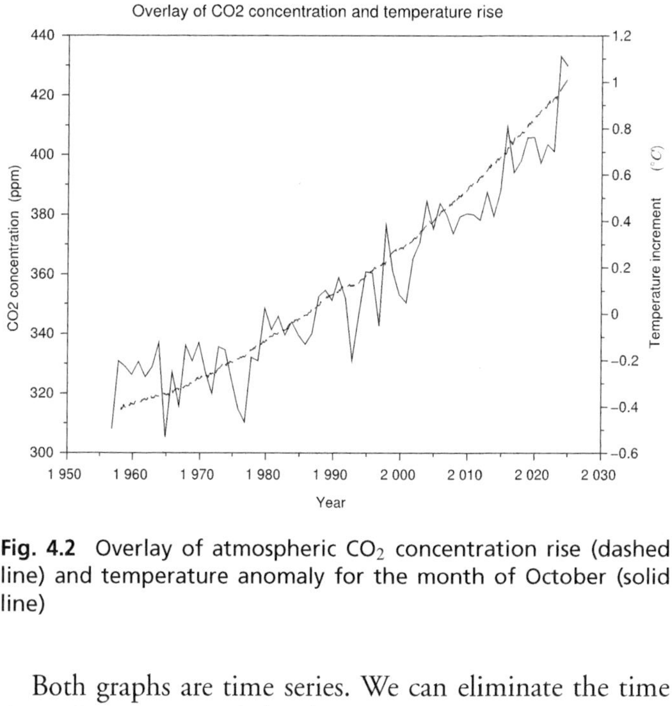

Fig. 4.2 Overlay of atmospheric CO_2 concentration rise (dashed line) and temperature anomaly for the month of October (solid line)

Both graphs are time series. We can eliminate the time dependency to reveal the direct relation between the two measured quantities. We do this by extracting pairs of concentration and temperature anomaly values that correspond to the same time. The result is the graph in Fig. 4.3, which suggests a roughly linear relationship between the CO_2 concentration and the temperature anomaly. For comparison, the red line shows a logarithmic dependency to be discussed later in Sect. 4.4.1.

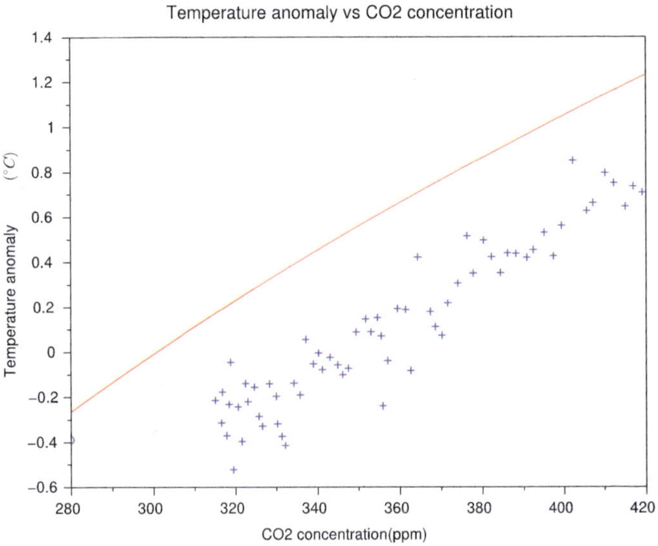

Fig. 4.3 Temperature anomaly vs atmospheric CO_2 concentration, both for the month of October (crosses). The solid curve represents a logarithmic increase with a slope similar to that of the data, starting at an initial temperature equal to the value assigned to the pre-industrial era

After observing the parallel trend of the CO_2 concentration and temperature anomaly, we face the central question of the global warming debate: Is this simply a correlation , or is one the cause of the other?[2] And, if there is a causal relation between the two, which is the cause and which is the effect? This question appears repeatedly in public

[2]We say that two variables are correlated when they happen to change in a coordinated way, but the causes of the changes are different and independent.

discussions about global warming. It is a fair question that is worth serious consideration.

Let us first consider the possibility that the rise in global mean temperature causes the CO_2 concentration to increase. The oceans contain large amounts of dissolved CO_2. Cold water can dissolve more CO_2 than warm water. When the oceans warm, they release CO_2 into the atmosphere because they can no longer hold all the CO_2 that was dissolved at lower temperatures. We are familiar with this phenomenon from fizzy drinks. A warm fizzy drink (beer or soda) foams over more easily when opened than a cold fizzy drink simply because a warm drink cannot dissolve as much CO_2 gas as a colder one. This dependency of the solubility of CO_2 on temperature supports the hypothesis that the warming of the Earth's surface (land and sea) causes an increase in CO_2 in the atmosphere. But then, why does the surface temperature of the Earth rise? We do not have a convincing answer to that. Some say it is because the climate is emerging from a cold period. That only kicks the can down the road in the search for a cause, because we still need to find out why the world is emerging from a cold period.

4.3.1 The 800,000-Year Temperature Record

Taking a closer look at how the Earth's temperature changed much further back in time may provide an answer. We can do this because we have a temperature record, shown in Fig. 4.4, spanning the last 800,000 years, obtained from the analysis of Antarctic ice cores.

Over the last million years, global temperature has gone up and down. One possible cause of these warming and cooling cycles is periodic changes in the Earth's orbit around the sun, which result in cyclic variations in the solar irradiance

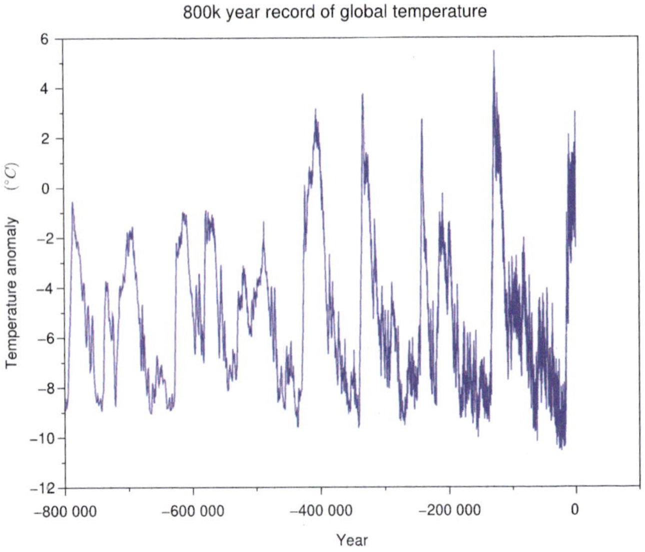

Fig. 4.4 Antarctic temperature anomaly for the last 800,000 years, relative to the average temperature of the last thousand years. Data from Jouzel et al. (2007)

reaching the Earth (Milankovitch cycles). The data indicate that since the last ice age, which ended approximately 10,000 years ago, Earth has been in a warm period, with temperatures fluctuating within ±2 °C of the last thousand-year average. Over that period, the solar irradiance due to the Milankovitch cycle has remained relatively unchanged. The consistent temperature rise observed over the last 150 years does not appear to be caused by changes in solar irradiance.

When we compare the 800,000-year Antarctic temperature record with the 800,000-year CO_2 concentration record extracted from the same Antarctic ice core, we see, in Fig. 4.5, that a strong correlation between the atmospheric CO_2 concentration and the temperature held over the last million years. Still, it leaves us no wiser as to which is the cause and which is the effect.

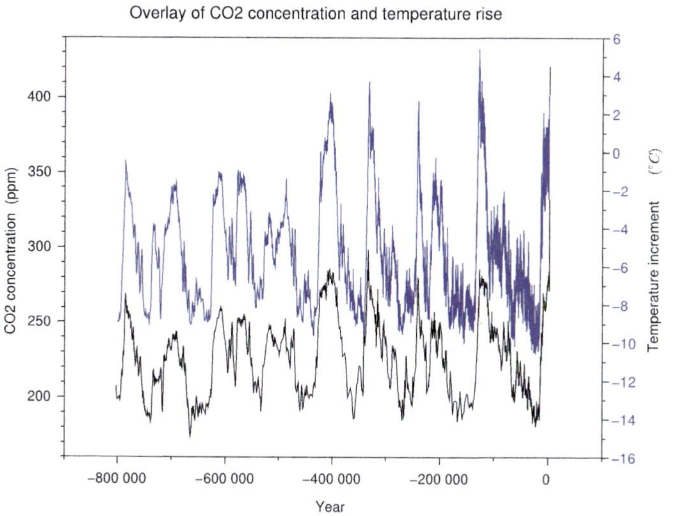

Fig. 4.5 Comparison of the 800,000 year history of the atmospheric CO_2 concentration (lower trace) and the Antarctic temperature (upper trace)

What we observe, though, is that the CO_2 concentration has shot up in recent times to levels not seen in the last 800,000 years, and that there is no corresponding temperature rise preceding this increase. The sudden increase in CO_2 concentration cannot be explained by CO_2 *outgassing* from a warming ocean. Instead, the evidence shows that, since the Industrial Revolution, humans have been releasing ever-increasing amounts of CO_2 into the atmosphere, and this release is responsible for the observed sudden, steep rise in atmospheric CO_2 concentration. Additionally, we know that an increase in atmospheric CO_2 concentration leads to a temperature rise due to the greenhouse effect, a correlation supported by observations. In the previous chapter, we analysed the mechanism for this effect. Hence, the observed facts strongly support the hypothesis that our CO_2 emissions are driving the current warming trend.

4.3.2 The 500 Million Year Climate Record

Ice cores only take us back about 800,000 years. Fossilised planktonic foraminifera can take us back 485 million years. Planktonic foraminifera are marine single-cell organisms that have a calcareous shell. The shells of dead planktonic foraminifera fall to the ocean floor, where they accumulate. This has been ongoing throughout the entire Phanerozoic eon. The Phanerozoic eon is the latest geological eon, ranging from 538.8 million years ago to the present.[3] The shells of the planktonic foraminifera grow from the chemicals available in the environment at the time of their formation. Therefore, the composition of the shells holds information about the conditions, including the climate, prevailing at the time the creatures lived. Fossilised planktonic foraminifera are abundant in ocean sediments, and their position in the strata of the sediment reveals their age. Through the analysis of fossilised planktonic foraminifera and other fossilised marine shells, it has been possible to roughly reconstruct the climate back to 485 million years (Judd et al. 2024).

There is a graph circulating on the internet showing an overlay of the reconstruction of the GMST and atmospheric CO_2 concentration over the last 570 million years, meant to demonstrate that there is no correlation between CO_2 concentration and temperature. Moore (2021) published the figure in his book, attributing it to Nasif Nahle, who references a 2002 article by C.R. Scotese. Other authors also published this graph, attributing it to MacRae (2008). MacRae is a journalist and blogger who published the graph on his website, giving no further reference than a geocraft.com website, where the trace ends. Despite the uncertain origin of the graph, the curves resemble the GMST reconstruction by

[3]The biggest time unit in the geological time scale is the *Eon*. Four eons cover the whole geological time scale.

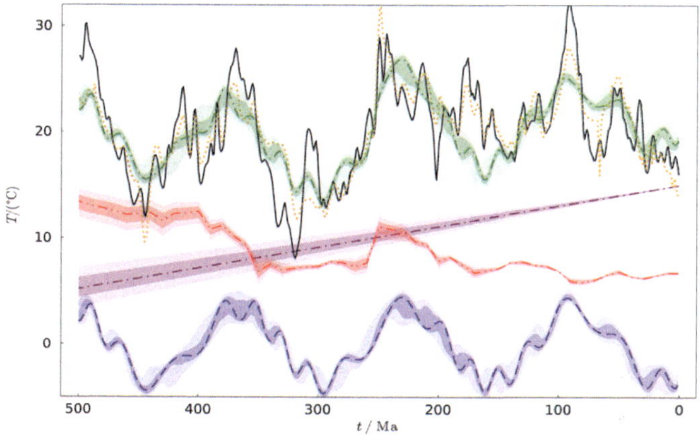

Fig. 4.6 Comparison of the 500 million year history of the atmospheric CO_2 concentration(dot-dot-dash) and reconstructed MGST temperature (solid). Figure from Shaviv et al. (2023)

C.R. Scotese and the atmospheric CO_2 concentration reconstruction by Berner and Kothavala (2001) around 2001. Since then, the reconstruction of GMST and atmospheric CO_2 over the Phanerozoic has improved (Royer et al. 2024; Scotese et al. 2021; Shaviv, Svensmark, and Veizer 2023) and, as in Fig. 4.6, shows significant correlation.

The most remarkable aspect of the current temperature rise is its rapid pace. Over the last 500 million years, temperature changes of a few degrees have taken millions of years. It took less than 200 years for a temperature rise of nearly 1.5 °C since the Industrial Revolution.

4.3.3 Changes in the Solar Irradiance

Any change in the total radiation energy emitted by the sun directly affects global temperature because it alters the Total Solar Irradiance (TSI) received by Earth. Solar irradiance changes are due to inherent solar processes and are separate from Milankovitch cycle effects. The best known of these changes is the 11-year sunspot cycle. Solar irradiance has changed minimally in recent history. TSI variation over the last solar cycle was around 0.1%. Indirect reconstruction of TSI over the past 9300 years shows that variation was less than 2 Wm^{-2} around the 1986 minimum of 1364.64 Wm^{-2} (Steinhilber, Beer, and Fröhlich 2009). Thus, over the last 10,000 years, solar irradiance remained constant within 0.15%.

With the advent of Earth satellites, it became possible to measure solar irradiance directly above Earth's atmosphere. Figure 4.7 shows the daily satellite measurements of solar irradiance since 1980.

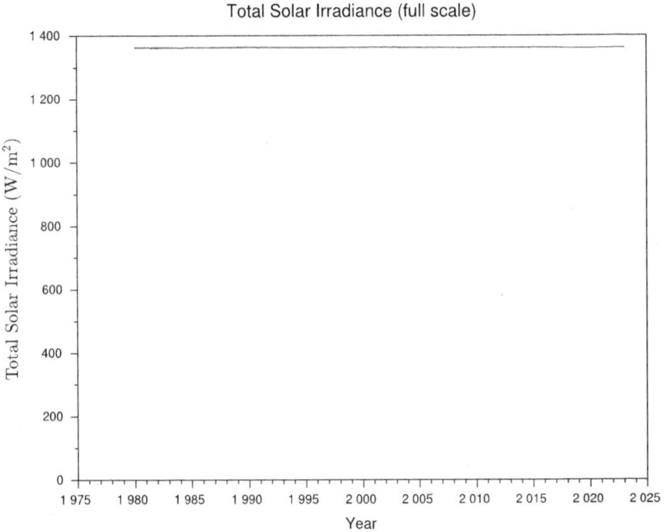

Fig. 4.7 Total Solar Irradiance (TSI) measured from satellites since 1980

The graph shows that solar irradiance has remained constant at around 1364.64 Wm^{-2} since direct measurements began. The slight variability of the solar irradiance arising from the 11-year sunspot cycle is too small to be visible at the scale of the plot in Fig. 4.7. To make the variations visible, we have to amplify the deviations around the mean, as in Fig. 4.8.

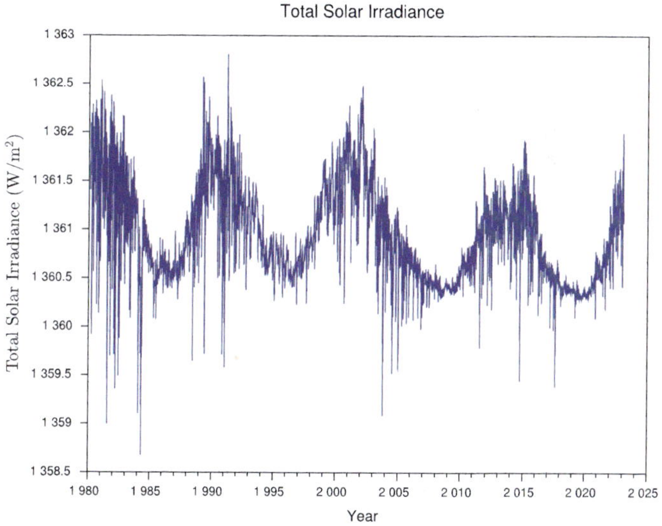

Fig. 4.8 Fine structure of the Total Solar Irradiance (TSI) measured from satellites since 1980

In this graph, the 11-year solar cycles are clearly visible. Still, the change in solar irradiation stays within a range of 2 Wm^{-2}, a very small fraction of the total irradiance. Small short-term fluctuations dominate the time series. These are inherently short-term irradiance fluctuations rather than measurement noise. Clearly, the sun is a bubbling cauldron. These measurements also indicate that since 1980, the solar irradiance averaged over the 11-year solar activity cycle has decreased by about 0.5 Wm^{-2}, while the mean surface temperature has risen by about 0.8 °C. It is not the sun that is frying us.

Changes in total solar irradiance over extended periods may be responsible for past climate variations. For example, the cold period known as the *Little Ice Age*,, which lasted from the 16th to the 18th century, coincided with the *Maunder* minimum in the number of sunspots observed between 1645 and 1715. Whether this period of diminished solar irradiance caused the Little Ice Age remains to be confirmed.

4.4 Temperature Rise Due to CO_2

In Sect. 3.5, we examined the Earth's radiation balance and concluded that greenhouse gases determine the mean surface temperature, acting like a thermostat. To see how temperature responds when we "turn the knob" by adding greenhouse gases, we next focus on CO_2, identified in Sect. 3.5.3 as the primary contributor to greenhouse warming. Thus, for now, we will disregard other greenhouse gases and attribute any temperature rise solely to CO_2.

Prominent 19th-century scientists–Fourier (1824), Tyndall (1862), and Arrhenius (1896) first proposed a connection between Earth's temperature and atmospheric CO_2. More than a century ago, Arrhenius (1986) estimated using available data that increasing CO_2 concentration by a factor of 3/2 would yield a 3 °C increase in temperature. He concluded that

> ...if the quantity of carbonic acid[4] increases in geometric progression, the augmentation of the temperature will increase nearly in arithmetic progression.

[4] In Arrhenius' time CO_2 used to be called carbonic acid.

4.4.1 Arrhenius' Calculations

Arrhenius calculated the mean surface temperature increase for various ratios of CO_2 concentration relative to a reference concentration. Specifically, he performed this calculation for ratios of 0.67 (2/3), 1.5 (3/2), 2, 2.5, and 3, for various latitudes and the four seasons. He observed that the calculated decrease in temperature from a reduction of CO_2 concentration by a ratio of 0.67 (2/3)[5] is about the same in magnitude as the calculated temperature increase from increasing the CO_2 concentration to a ratio of 1.5 (3/2). The calculated temperature changes range from 2.9 to 3.7 °C, with a predominantly 3 °C increase at the equator for all seasons. From Arrhenius's calculations, we can formulate the simplified rule that when the CO_2 concentration increases from an initial CO_2 concentration c_0 to $3/2 \cdot c_0$, the temperature increments by 3 °C.

$$\Delta T = T\left(\frac{3}{2} \cdot c_0\right) - T(c_0) = 3 \qquad (4.1)$$

Pre-industrial CO_2 concentration was about 280 ppm, rising to around 420 ppm by the end of 2022. Interestingly, $420 = 280 \cdot 3/2$, so, following Arrhenius' calculation, the temperature should have risen by just over 3 °C (see Mathbox 4.4.1).

[5]Note that $0.67 \cdot (3/2) = 1$ and $1 \cdot 3/2 = 1.5$).

Mathbox 4.4.1

Arrhenius formula for CO_2 induced warming

According to the Arrhenius formula 4.1

$$\Delta T = T\left(\frac{3}{2} \cdot c_0\right) - T(c_0) = 3 \quad (4.2)$$

A further rise in the concentration from $3/2 \cdot c_0$ by another factor of 3/2 to $(3/2) \cdot (3/2)c_0 = (3/2)^2 \cdot c_0$ rises the temperature again by 3 °C. We can write this in the following table:

x	Concentration c(x)	Temperature increase (°C)
0	$(3/2)^0 \cdot c_0$	$0 \cdot 3 = 0$
1	$(3/2)^1 \cdot c_0$	$1 \cdot 3 = 3$
2	$(3/2)^2 \cdot c_0$	$2 \cdot 3 = 6$
3	$(3/2)^3 \cdot c_0$	$3 \cdot 3 = 9$

The temperature rises linearly as the concentration grows exponentially in x. $c(x) = \left(\frac{3}{2}\right)^x \cdot c_0$ (4.3)

We take the logarithm on each side to solve for x

$$\ln\left(\frac{c(x)}{c_0}\right) = x \cdot \ln\left(\frac{3}{2}\right) \quad (4.4)$$

$$x = \frac{1}{\ln\left(\frac{3}{2}\right)} \cdot \ln\left(\frac{c(x)}{c_0}\right) \quad (4.5)$$

The temperature rise ΔT is $3 \cdot x$, therefore

$$\Delta T = 3 \cdot x = 3 \cdot \frac{1}{\ln\left(\frac{3}{2}\right)} \cdot \ln\left(\frac{c(x)}{c_0}\right) = 7.35 \cdot \ln\left(\frac{c(x)}{c_0}\right) \quad (4.6)$$

For the increase of CO_2 concentration from 280 ppm in pre-industrial times to 420 ppm in 2022, the temperature increase is

$$\Delta T\left(\frac{420}{280}\right) = 3 \cdot \frac{1}{\ln(\frac{3}{2})} \cdot \ln\left(\frac{420}{280}\right) = 3 \cdot \frac{1}{\ln(\frac{3}{2})} \cdot \ln\left(\frac{3}{2}\right) = 3 \quad (4.7)$$

degrees C.

However, the observed temperature increase from the pre-industrial era to 2022 was about 1.1 °C. This indicates that Arrhenius' calculations overestimated the temperature increase. Even so, his manual calculation is an outstanding example of using fundamental physics to answer a difficult question, almost as well as today's climate scientists do, with the help of large and complex numerical models.

Since Arrhenius' time, extensive and detailed measurements of climate variables have enabled us to refine our understanding of the atmosphere's role in balancing incoming solar radiation energy. Moreover, detailed mathematical modelling of the radiative transfer processes (Myhre et al. 1998) due to greenhouse gases, which takes into account all the individual absorption lines, gives a logarithmic dependence of the radiative forcing ΔF_n by CO_2 on the CO_2 relative concentration ratio:

$$\Delta F_n = 5.35 \cdot \ln\left(\frac{c(x)}{c_0}\right) \tag{4.8}$$

As we shall find in Sect. 4.6, calculating the global mean surface temperature change from radiative forcing is quite laborious. Despite this, it is important to note that the temperature change is proportional to the radiative forcing, so the temperature change is also proportional to the logarithm of the concentration ratio c/c_0–just as Arrhenius found.

The logarithmic dependence is conspicuously visible in the observed dependence of temperature on the CO_2 concentration displayed in Fig. 4.3 because the logarithmic curve (red line) is almost a straight line over the relatively small concentration range for which we have data.

4.4.2 Extrapolating the Recent Trend in GMST Rise

We can make a practical estimate of the temperature increase per Gigatonne of CO_2 added to the atmosphere using available observational data, without relying on climate models. By analysing the Keeling curve and NOAA temperature data from 1974 to 2024, when the temperature increase appears nearly linear, we find a rate of 0.023 °C per year, totalling 1.173 °C over 51 years. To link this observed temperature rise directly to increased atmospheric CO_2, we need to determine the change in CO_2 concentration over these 51 years. The Keeling curve data set shows that at the beginning of 1974, the atmospheric CO_2 concentration was 329.45 ppm. At the end of 2024, the measured concentration was 425.87 ppm. This is a 96.42 ppm increase over 51 years. Dividing the temperature rise by the increase in CO_2 concentration reveals how much the temperature rises with each one part per million increase in CO_2.

$$\frac{\Delta T}{\Delta c} = \frac{1.173\ ^\circ\mathrm{C}}{96.42\ \mathrm{ppm}} = 0.01217 \frac{^\circ\mathrm{C}}{\mathrm{ppm}} \tag{4.9}$$

To give this result more practical meaning, we should consider how much CO_2, by mass, produces a 1 ppm change in atmospheric concentration–since such a mass-based figure is more tangible than molecular concentration.

To do this conversion, we must translate the change in molecular concentration to the mass of CO_2 released into the atmosphere. We already calculated this conversion factor in Chap. 2. We found that, for small concentrations, the CO_2 weight concentration is 1.53 times the CO_2 molecular concentration. To find the total mass of CO_2, multiply the per mass concentration change by the mass of the atmosphere. In Chap. 2, we already calculated the mass of the atmosphere to be 5.268 10^{15} tonnes.[6] The mass of carbon in a mole of CO_2 is 12/44 = 0.27 times the mass of CO_2. Then, according to our calculation, one ppm molecular concentration of CO_2 in the atmosphere is

$$5.268\ 10^{15}\ \text{tonnes} \cdot 1.53 \cdot 0.27\ 10^{-6} = 2.176\ 10^{9}\ \text{tonnes} \tag{4.10}$$

of carbon. The difference with the value reported in the literature (Friedlingstein 2020) is 2.4%, which is likely due to a discrepancy in the estimate of Earth's atmospheric mass. We can now translate the increase of molecular CO_2 concentration by 96.42 ppm over the 51 years from 1974 to 2024 to the corresponding increase of the mass of CO_2 in the atmosphere:

$$\begin{aligned} 5.268\ 10^{15}\ \text{tonnes} \quad &\cdot 1.53 \cdot 96.42\ 10^{-6} = 777.15\ 10^{9}\ \text{tonnes} \\ &= 777.15\ \text{Gt} \end{aligned} \tag{4.11}$$

The last step in our calculation for finding the temperature increase per mass of CO_2 increase is to divide the temperature increase during these 51 years, as calculated above, by the increase in the mass of CO_2 over the same period

$$\frac{\Delta T}{\Delta M_{CO_2}} = \frac{1.173\ °\text{C}}{777.15\ \text{Gt}} = 0.00151 \frac{°\text{C}}{\text{Gt}} \tag{4.12}$$

Thus, the release of one Gt of CO_2 into the atmosphere will raise the mean global surface temperature by 0.00151 °C, if the temperature increase is attributed exclusively to the rise in CO_2 concentration.

The IPCC does this calculation in a slightly different way. The IPCC defines the Equilibrium Climate Sensitivity

[6]In the literature, atmospheric CO_2 concentrations are often converted directly from molecular concentration in parts per million to petagrams ($1Pg = 1 \cdot 10^{15}\ \text{g} = 1\text{Gt}$) of carbon per ppm of CO_2 in the atmosphere by using the conversion factor 2.124 PgC/ppm (Friedlingstein 2020).

(ECS) to express their estimate of the temperature increase due to CO_2 accumulation in the atmosphere. ECS is the equilibrium (steady state) change in the surface temperature following a doubling of the atmospheric CO_2 concentration from the pre-industrial level, without any further emissions. The IPCC best estimate for ECS is 3 °C.

Let us compare their estimate with the result of our linear extrapolation of the trend of the last 51 years. Pre-industrial atmospheric concentration of CO_2 was 280 ppm. Using the factor calculated in Eq. 4.11, duplicating the pre-industrial CO_2 concentration of 280 ppm gives a temperature increase of

$$0.01217\,\frac{°\mathrm{C}}{\mathrm{ppm}} \cdot 280\ \mathrm{ppm} = 3.41\ °\mathrm{C} \tag{4.13}$$

Extrapolation of the trend of the last 51 years gives an Equilibrium Climate Sensitivity of 3.41 °C. The relative difference between our calculation and the IPCC best estimate is

$$\frac{3.41\ °\mathrm{C} - 3.0\,°\mathrm{C}}{3.0\,°\mathrm{C}} = 0.136 \tag{4.14}$$

Our basic, back-of-the-envelope calculation differs from the IPCC best estimate by only 14%. Not too bad!

4.5 The CO_2 Emission Budget

For the temperature rise predicted by the extrapolation in Fig. 4.1 to hold, future yearly CO_2 emissions must stay close to their average over the past 51 years, since 1974. However, there is no certainty that CO_2 emissions will remain at this level. We cannot predict exactly how much CO_2 will be emitted in the future, but we do know the expected temperature rise for any given amount of emitted CO_2. This allows us to calculate how much additional CO_2 is required to produce a specified temperature increase, regardless of the time frame. In other words, there is a CO_2 emissions budget. For instance, it is possible to determine how much CO_2 can be emitted before warming reaches 1.5 °C above pre-industrial levels, which is the target set in the 2015 Paris Accord. How quickly we reach that limit depends on how we manage emissions.

According to the NOAA temperature anomaly data, the 2024 mean surface temperature was 0.9788 °C above the 20th century mean and 1.2488°C above the pre-industrial level, which is 0.2512 °C away from the limit. We can immediately calculate how much additional CO_2 will produce a 0.2512 °C temperature increase by dividing this increase by the temperature increase per Gt that we calculated above

$$\frac{0.2512\,^{\circ}\mathrm{C}}{1.51\ 10^{-3}\,\frac{^{\circ}\mathrm{C}}{\mathrm{Gt}}} = 166\ \mathrm{Gt} \tag{4.15}$$

The allowed *budget* for atmospheric CO_2 accumulation before the mean surface temperature rise reaches 1.5 °C is alarmingly small when compared to the 776.58 Gt added in the past 51 years. Our budget is only one-fifth of what was added to the atmosphere in that time, underscoring the urgency: at the current accumulation rate, our budget will be exhausted, at the latest, in $51 \cdot 166/776.58 = 10.9$ years from 2024; that is, by 2035.

Since 1974, 776.58 Gt of CO_2 have been added to the atmosphere. This averages 15.23 Gt per year. The Keeling curve (Fig. 4.2) bends upward, showing that the current rate is above average. Our budget may run out before 2035 unless emissions slow. The rise in temperature will slow slightly as CO_2 concentration increases, since the relationship is logarithmic. This may compensate somewhat for higher emissions.

4.5.1 Natural CO_2 Removal

The calculated yearly increase of about 15 Gt of CO_2 in the atmosphere, implied by the Keeling data, is much lower than the estimated 40 Gt of CO_2 released in 2023 from fossil fuel burning and other sources. This discrepancy raises the question: what happens to the apparently missing 25 Gt?

The answer lies in natural processes that remove CO_2 from the atmosphere through photosynthesis in plants and by dissolving into the oceans. Other natural processes release CO_2, including animal and plant respiration, decomposition of organic matter, ocean outgassing, volcanic activity, and wildfires. So, while CO_2 emissions add to the atmosphere, some is removed by the carbon cycle.

It is essential to understand the size of these natural fluxes. Estimating the mass of CO_2 in this cycle is not easy. Estimates range from 10 to 100 Gt per year (Riebeek, 2011). This is many times more than what humans emit. However, more important than the size is that over hundreds of thousands of years, the atmospheric CO_2 concentration has fluctuated around a stable mean.

The geological record of CO_2 (see Fig. 1.2) shows that, before large-scale human emissions, natural processes kept long-term CO_2 levels in balance. That balance no longer exists. Natural processes cannot remove all of the extra CO_2 from human activities, so atmospheric concentrations keep rising.

We can do another simple calculation to find what fraction of CO_2 from human activity remains in the atmosphere. For recent years, we know how much CO_2 has been put into the atmosphere by burning fossil fuels. Using data from the BP Statistical Review of World Energy, which goes back to 1958, we can estimate the mass of CO_2 emitted from burning fossil fuels since 1960. The Keeling curve data now allows us to estimate the annual increase in atmospheric

CO_2. Figure 4.9 compares the mass of CO_2 from fossil fuel emissions with the net increase in atmospheric CO_2 calculated from the Keeling curve.

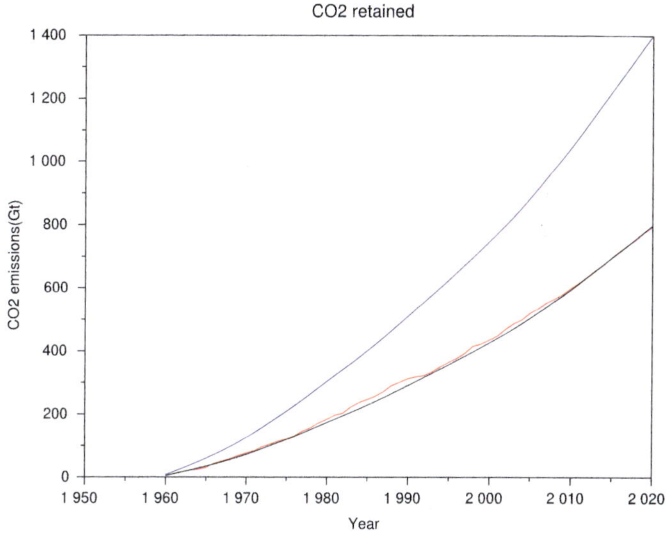

Fig. 4.9 CO_2 emission into the atmosphere from burning fossil fuels (upper line) from 1960 to 2020, compared to the increment of CO_2 content in the atmosphere (lower line). The dashed line is 57% of the global fossil fuel emission curve

The increase in the mass of CO_2 in the atmosphere (red line) closely matches 57% of fossil fuel emissions (black line). It appears that natural processes remove 43% of the emitted CO_2, leaving 57% in the atmosphere.

Burning fossil fuels is not the only human activity that releases CO_2 into the atmosphere. Deforestation and other forms of vegetation removal emit significant amounts of CO_2 into the atmosphere. The technical literature refers to these CO_2 emissions as arising from *land use change.* These are difficult to quantify and exhibit significant variability.

Friedenberg et al. estimate that CO_2 emissions from land-use change in 2019 were around 6.6 Gt/year. According to data from the BP Statistical Review of World Energy, in 2019, the emission from burning fossil fuels was 34.357 Gt. Adding the land-use change emissions to the fossil fuel emissions yields 40.957 Gt of anthropogenic CO_2 emissions in 2019. Thus, land-use emissions add 19.2% of the fossil fuel emissions. Let us assume that this proportion has remained constant over the last 60 years. Then we have to multiply the emission curve in Fig. 4.9 (blue line) by 1.19 to get the total anthropogenic CO_2 emissions, and the amount retained in the atmosphere falls from 57% to 48% ($0.57/1.19 = 0.479$).

4.5.2 The Revised CO_2 Emission Budget

Because natural processes remove 52% of CO_2 emissions from the atmosphere, we must distinguish between how much we emit and how much actually stays in the air. The amount emitted is more than twice the amount that remains to increase the CO_2 concentration. With this in mind, the budget for CO_2 increase in the atmosphere becomes

$$\frac{166 \text{ Gt}}{0.479} = 346.6 \text{ Gt} \tag{4.16}$$

for emission of CO_2 by human activities. The larger CO_2 budget does not change how long the budget will last. If fossil fuel burning and land clearing continue at about 40.95 Gt/year as in 2019 (with over 40 Gt estimated in 2023), the budget will last 8.46 years from 2024 (346.6/40.95). After 2033, each tonne of accumulated CO_2 will push temperatures past 1.5 °C above pre-industrial levels.

We can extend that *grace* period by reducing emissions gradually from now on. If we lower our CO_2 emissions by a constant amount every year, reducing them linearly to zero, we can make the budget last twice as long.[7]

The total mass of CO_2 in the atmosphere at the end of 2024 was

$$\begin{aligned} &5.268\ 10^{12}\ \text{tonnes} \cdot 1.53 \cdot 425.87\ 10^{-6} \\ &= 3433\ 10^{9}\ \text{tonnes} \end{aligned} \tag{4.17}$$

Our annual emissions of 41 Gt are just 1.2% of that. Removing 52% of the yearly emissions into the atmosphere amounts to

$$41\ 10^{9}\frac{\text{Gt}}{\text{y}} * 0.52 = 21.3\ 10^{9}\ \frac{\text{Gt}}{\text{y}} \tag{4.18}$$

tonnes per year. From pre-industrial times to 2024, CO_2 concentration rose from 280 ppm to 425.87 ppm. This corresponds to a mass of

$$\begin{aligned} &5.268\ 10^{15}\ \text{tonnes} \cdot 1.53 \cdot (425.87 - 280)\ 10^{-6} \\ &= 1176\ 10^{9}\ \text{tonnes} \end{aligned} \tag{4.19}$$

of CO_2 added to the atmosphere by human activity since the pre-industrial era.

If we stopped all CO_2 emissions in 2024, natural processes, acting at 21.3 Gt per year, would take

$$\frac{1176/2\ 10^{9}\ \text{Gt}}{21.3\ 10^{9}\frac{\text{Gt}}{\text{y}}} = 27.6\,\text{y} \tag{4.20}$$

[7]The average of a linear reduction is half the initial value, hence $3(46.6/2\,\text{Gt})/40.95\frac{\text{Gt}}{\text{y}} = 17\,\text{y}$. If we started in 2024, we would have until 2041 before reaching the CO_2 emissions limit.

return the atmosphere to pre-industrial CO_2 levels. Eventually, the global mean surface temperature would return to pre-industrial levels.

Notice that the goal of *net zero* CO_2 by 2050 means that only as much CO_2 is added to the atmosphere as is removed by natural processes, resulting in a stable CO_2 concentration in the atmosphere, and therefore no further increase in temperature, but no decrease in temperature either.

4.6 Contribution to Warming by Each Greenhouse Gas

Our calculations so far of the CO_2 budget are simple but flawed. They blame all warming on increased atmospheric CO_2. In reality, all greenhouse gas emissions drive the actual warming (Fig. 4.1). To estimate a CO_2 budget for a temperature limit, we must separate CO_2's impact from other gases. Because individual contributions can't be measured directly, we rely on model-based estimates.

In the previous chapter, we examined the Earth's energy balance by examining what happens to the incoming radiation flux from the Sun. In Sect. 3.6, we examined the notion of *radiative forcing* used by climate scientists to quantify the changes in the net radiation fluxes that determine the Earth's energy balance. NOAA calculates estimates of the radiative forcing from individual greenhouse gases. From these estimates, it is possible to calculate the amount of energy retained due to each gas. To do so, we must integrate the change in flux over the entire surface of the Earth and over time to obtain the net change in energy retained by the Earth during that period.

Still, the amount of retained energy does not tell us how much the mean surface temperature will rise due to that additional energy. The *specific heat* c_p of a material tells us the amount of thermal energy ΔQ required to raise the temperature of a mass m of material by an amount ΔT, at constant pressure

$$\Delta Q = m \cdot c_p \Delta T \tag{4.21}$$

or

$$\Delta T = \frac{\Delta Q}{m \cdot c_p} \tag{4.22}$$

Earth's surface materials absorb thermal energy and warm up according to their specific heat. Estimating the temperature rise is complex because each material absorbs different amounts of energy to reach atmospheric temperature. Calculating a total temperature increase requires summing the energy absorbed by all surface materials, taking into account their specific heats.

We can, however, complete one part of the calculation, which is the one for the oceans. At the surface of the oceans, the atmosphere is in contact with only one substance: water. How much energy does it take to raise the ocean's surface layer, say the first 10m, by one degree? The calculation will illustrate the oceans' contribution to global temperature rise. Oceans cover 71% of the Earth's surface. The global ocean surface area is 361 10^6 km^2.

The volume of water in a 10 m-thick layer is

$$V = 361\ 10^6\ \text{km}^2 \cdot 10\,\text{m} = 361\ 10^{13}\,\text{m}^3 \tag{4.23}$$

At a density of 1000 kg/m^3 this is a mass of

$$M = \rho \cdot V = 361\ 10^{13}\,\text{m}^3 \cdot 1000\,\frac{\text{kg}}{\text{m}^3} = 361\ 10^{16}\,\text{kg} \tag{4.24}$$

The specific heat of water is $c_p = 4.184$ kJ/kg °C. Therefore

$$\Delta Q = m \cdot c_p \Delta T = 361\ 10^{16}\,\text{kg} \cdot 4.184\,\frac{\text{kJ}}{\text{kg °C}} \cdot 1\,°\text{C}$$
$$= 1510\ 10^{19}\,\text{J} = 15{,}100\,\text{EJ} \tag{4.25}$$

A surface layer of 10 m of ocean water can absorb 15,100 EJ before its temperature increases by 1 °C. How does this compare with the total amount of excess energy caused by total emissions since the industrial era began?

There is a way to simplify the calculation of the temperature increase resulting from the additional energy retained in the atmosphere. It rests on recognising that the extra energy supplied to the Earth per unit time is proportional to radiative forcing ΔF_n. Thus, using Eq. 4.22, we can say that the temperature increase is proportional to the total radiative forcing

$$\Delta T = \lambda\ \Delta F_n \tag{4.26}$$

and λ can now be found from the ΔF_n and the *observed* temperature change. The estimate of the total radiative forcing since the beginning of the industrial era is 2.72 W/m^2

However, we still face the same problem: the temperature change we observe is the result of all contributors, and any change measured in the outgoing radiative flux is a result of all contributors. We can only single out the contribution to global warming from each greenhouse gas by calculating the fraction of the total ΔF_n coming from each greenhouse gas.

Expressing the contribution of each gas in multiples of the contribution of a reference gas makes this calculation easier. The chosen reference gas is CO_2 because it is the dominant gas for atmospheric warming, and we express the contribution of other greenhouse gases in terms of CO_2 *equivalent* emissions. The CO_2 equivalent (CO_2-eq) of a particular greenhouse gas, as the name says, is the mass of CO_2 that would produce the same amount of atmospheric heating as a given amount of that greenhouse gas. The World Meteorological Organisation published values of *Global Warming Potential* for the greenhouse gases (Burkholder, Hodnebrog, and Orkin 2018). The GWP-100, which indicates the total contribution over 100 years, has been widely used in climate studies to report emissions of different greenhouse gases on a common scale. The GWP-100 for methane is around 28. That means that cumulative warming over 100 years from the emission of one kilogram of methane is the same as the emission of 28kg of CO_2. Because CO_2 is the reference gas, its GWP-100 is, of course, equal to one.

Now, we can add the CO_2-eq emissions from all greenhouse gases and express the combined contribution in a single amount of CO_2-eq emissions. In this way, our carbon budget, previously based on attributing all warming solely to CO_2 emissions, effectively becomes a budget for total CO_2-eq emissions from all greenhouse gases. The CO_2-only emission budget is now significantly smaller, reflecting the need to also account for the contributions of other gases when comparing allowable future emissions.

4.6.1 The Methane Contribution

After CO_2, methane is the next most important greenhouse gas. Reports in the media often highlight methane's large contribution to global warming despite its low concentration. They quote values of up to one-third for the methane contribution, without providing a proper reference for these estimates. Environmental advocacy groups use the supposed large contribution of methane to promote policies that could have drastic consequences for the livestock industries and pressure governments to implement policies to reduce methane emissions. In these circumstances, we need to assess the contribution of methane emissions to global warming ourselves.

The GWP-100 of methane is 28. That means one kilogram of methane contributes 28 times as much to global warming as one kg of CO_2. To estimate methane's warming contribution, we need to find the weight of the methane in the atmosphere. We first need to convert the molecular concentration to weight concentration as we did for CO_2.

NOAA has measured the methane concentration in the atmosphere since 1983. From the end of 1984 to the end of 2022, the methane concentration has increased from 1644.69 ppb (parts per billion) to 1911.84 ppb. By the end of 2021, the mean atmospheric molecular concentration of methane was 1,895.29 parts per billion (ppb). The molecular weight ratio of methane (CH_4) to air is $16/28.8 = 0.556$. A given number of CH_4 molecules weighs 0.556 times as much as the same number of *air* molecules. If we now take away a fraction c of molecules from a mole of air mixture molecules and replace them with the same fraction of CH_4 molecules to leave the total number of molecules the same, then the total mass of one mole of the mixture is

$$\begin{aligned} m &= (1-c)\cdot 28.8\,\mathrm{g} + c\cdot 16\,\mathrm{g} = 28.8\,\mathrm{g} + c\cdot(28.8\,\mathrm{g} - 16\,\mathrm{g}) \\ &= 28.8\,\mathrm{g} - c\cdot 12.8\,\mathrm{g} \end{aligned} \tag{4.27}$$

The weight fraction of CH_4 is

$$\frac{m_{CH_4}}{m} = \frac{c\cdot 16}{28.8 - c\cdot 12.8} \tag{4.28}$$

If c is very small then $28.8 - c\cdot 12.8 \approx 28.8$ and the per weight CH_4 concentration differs insignificantly from the molecular weight ratio

$$\frac{m_{CH_4}}{m} \approx c\frac{16}{28.8} = 0.556\cdot c \tag{4.29}$$

To get the total mass of methane in the atmosphere, we multiply the mass of the atmosphere, which we calculated earlier to be $5.268\ 10^{15}$ tonne, by the mass concentration of methane $0.556 \cdot 1895.29\ 10^{-9}$

$$5.268\ 10^{15}\,\text{tonne} \cdot 0.556 \cdot 1895.29\ 10^{-9} = 5550\ 10^{6}\,\text{tonne} = 5.6\,\text{Gt} \tag{4.30}$$

We calculated the mass of CO_2 in the atmosphere to be 3364 Gt in 2021. Thus, with a GWP-100 of 28, the atmosphere's methane mass contributes as much as $28 \cdot 5.550\,\text{Gt} = 155\,\text{Gt}$ of CO_2. In relative terms, methane contributes $155/3364 = 0.46.2$, or 4.6%, of the warming produced by the CO_2 in the atmosphere.[8]

Methane molecules absorb more infrared radiation than CO_2 molecules, but have a much shorter life in the atmosphere than CO_2 molecules. The GWP-20 of methane is 84 (Burkholder et al. 2018), nearly three times its GWP-100 of 28. The GWP-100 is much smaller than the GWP-20 because, after 20 years, only a small portion of the original methane remains in the atmosphere. In estimating the contribution of methane over 20 years relative to CO_2, we count all the methane's contribution and only a fifth of the hundred-year contribution of CO_2. Multiplying the methane weight concentration by its GWP-20 gives $1053.78\ 10^{-9} \cdot 84 = 88.518\ 10^{-6}$ equivalent of CO_2. The warming contribution of methane is $88.518\ 10^{-6}/638.64\ 10^{-6} = 0.139$ times that of the CO_2 in the atmosphere.

The International Energy Agency (IEA)[9] states that methane contributed 30% to global warming. They give

[8]Without multiplying by the mass of the atmosphere, we can get the same result by multiplying the weight concentration of methane by 28 and then take the ratio of the result to the CO_2 weight concentration.

[9]https://www.iea.org/reports/global-methane-tracker-2022/methane-and-climate-change.

no source for that estimate. The UN Environment Program (UNEP) keeps quoting this statement without evidence.[10] The UNEP further states that over 20 years methane produces 80% more warming than CO_2. Clearly, this is incorrect. The GWP-20 of methane is 83 on an equal mass basis. This value must be multiplied by the much lower atmospheric methane mass to obtain the actual contribution, as we did above. It further ignores that CO_2 contributes to warming for a much longer period. Hence, the GWP-100 factor of methane is 28. Is there an ideologically driven misrepresentation of the data by the UNEP?

The UNEP makes the following misleading statement:

> Carbon dioxide remains in the atmosphere for hundreds to thousands of years. Therefore, even reducing emissions immediately and dramatically would not affect the climate until later in the century. But it takes only about a decade for methane to break down. Therefore, reducing methane emissions now would have an immediate impact and is crucial for helping the world stay on a path to 1.5 °C.

Saying that an immediate halt to CO_2 emissions would not affect the climate is misleading. Clearly, it would stop the temperature rise. If all methane emissions were to stop now, then after 20 years (a 12-year lifetime), very little methane would remain in the atmosphere, and the warming contribution from methane would have vanished. Only the atmosphere's CO_2 would control the temperature. Until then, the temperature increase might slow down as the methane contribution fades. Stopping all human-sourced methane emissions would result in only an 8% reduction of the temperature rise from CO_2-eq emissions. Stopping all methane emissions is overly optimistic, because only those from human

[10] https://www.unep.org/news-and-stories/story/methane-emissions-are-driving-climate-change-heres-how-reduce-them.

activity, estimated at 60% of total emissions, can be stopped. The remaining emissions from natural sources would continue.

Our calculations show that human-sourced methane emissions contribute to global warming, but much less than some interest groups claim. Reducing methane emissions is much less effective than reducing CO_2 emissions. Burning methane rather than releasing it into the atmosphere reduces methane emissions. To see why, consider burning 1 kg of methane. Burning 1 kg of methane releases 2.44 kg of CO_2. In this case, the CO_2-eq of methane reduces from 28 to just 2.44. However, we must add the warming from the heat released when burning methane.

Burning 1 mole of methane (16 g) releases 891 kJ. Burning 1kg of methane releases $(1000/16) \cdot 891\,\text{kJ} = 55.7\text{GJ}$ of thermal energy. To compare this with the effect of releasing 2.44 kg of CO_2 into the atmosphere, we need to know how much radiation energy 2.44 kg of CO_2 absorbs over 100 years in absolute terms. That is when things become complicated, and we must approach this question from a different angle.

According to the energy balance data analysed in Sect. 3.4, the average incoming energy flux from the sun absorbed by the Earth's surface is $163.3\,\text{W/m}^2$. For one year, the total over the Earth's surface is

$$
\begin{aligned}
&163.3\,\frac{\text{W}}{\text{m}^2} \cdot 510.06\;10^{12}\,\text{m}^2 \cdot 31.536\;10^6\,\frac{\text{s}}{\text{y}} \\
&= 2.63\;10^{24}\,\frac{\text{J}}{\text{y}} = 2.63\;10^6\,\frac{\text{EJ}}{\text{y}} \qquad (4.31)
\end{aligned}
$$

Assuming that 80% of fossil fuel energy consumption ends up as waste heat released into the atmosphere, then the waste heat released into the atmosphere in 2023 from fossil fuel consumption was

$$504.83 \text{ EJ} \cdot 0.8 = 403.9 \text{ EJ} \tag{4.32}$$

The average waste energy released into the atmosphere from fossil fuel consumption per year is $404/2.63\ 10^6 = 0.00015$ of the solar energy absorbed by the Earth's surface. That is 0.01545%. From this, we can conclude that the combustion energy from burning methane makes a negligible contribution to global warming.

4.6.2 Human and Animal Respiration

Animal respiration emits CO_2 into the atmosphere. How much does our respiration contribute to global warming?

In one day, the average person exhales around 500 l of the greenhouse gas CO_2—which amounts to approximately 1kg in mass. (See Wikipedia: Respiratory system for details) Multiplied by the world population of 8 billion, this is $8\ 10^9$ kg/day = 8 Mt/day. In a year, this is $8 \text{ Mt} \times 365 = 2.92$ Gt and would be a significant contribution. However, this CO_2 is part of the carbon cycle and comes from the carbohydrates we ingest as food. Plants produce these carbohydrates by photosynthesis from atmospheric CO_2. Therefore, human and all animal breathing makes no net contribution to the atmosphere's CO_2 content.

4.7 The Weather

So far, we have analysed the *average* rate of climate warming. However, the local deviations from the average are what create the weather. Local deviations arise from the variation of incident solar flux from the equatorial regions to the poles. The incident solar flux diminishes as we move from the equator to the poles. The absorption and reflection of solar radiation differ between land and sea. On land, it also varies from place to place, depending on local conditions.

Consequently, surface temperatures are far from uniform across the planet. These temperature differences cause the atmosphere to heat differently. Areas of higher temperature create warmer air near the surface, which then rises, producing air currents. Warm air absorbs moisture, which it releases as rain and snow when it cools. Aided by the inertial forces arising from the Earth's rotation, the atmosphere becomes a turbulent air mass that swirls around the globe, making *the weather*. Global warming mostly manifests as changes in the weather. Warmer air holds more water vapor, leading to increased rainfall in some regions. Bigger temperature differences create stronger air currents (storms). Established air and sea currents can change direction and strength, which in turn affects rainfall patterns. Many phenomena contribute to weather, making prediction hard. We can say with certainty that climate warming will intensify weather events. The danger lies in the long-term unpredictability of extreme weather (storms, floods, heat waves, and droughts). Not knowing when or where they will occur makes preparation difficult, so they often have severe impacts.

4.8 The Bottom Line

Since the industrial revolution,, we have already dumped enough CO_2 into the atmosphere to raise the temperature to a level that endangers not only economic growth but humanity's way of life itself. The current way of producing and using energy has reached a point where it limits the exponential growth of the world economies we have become used to. To stop further climate warming, we must immediately stop pouring CO_2 into the atmosphere. As of today, we should not burn any fossil fuel to generate energy!

If we completely stopped burning fossil fuels, we would have to live with the 15.7% of our current total energy consumption that we can generate by other means (nuclear, hydro, wind, and solar). Given the reliance of our economies on energy, we are very reluctant to do that. Hence, the defining issue for the world economy over the next decades is replacing the 80+% of our energy consumption from fossil fuels with energy sources that do not generate CO_2 pollution. Replacing fossil fuels as an energy source is a big task, considering that we have not yet solved much simpler problems, such as preventing deadly amounts of plastic waste from entering the environment.

5

Fossil Fuels

From the beginning of the Industrial Age to today, fossil fuels have provided the energy necessary to create and sustain the affluent lifestyle available to the inhabitants of wealthier nations on our planet. In 2023, fossil fuels provided 81.5% of the energy needed to power our industrial economies. Fossil fuels remain abundant and affordable. There seemed to be no limits to their use until, in the early twenty-first century, it became clear beyond a doubt that the Earth's atmosphere is warming because of the large amount of CO_2 emitted by burning fossil fuels.

To prevent further global warming, dumping CO_2 into the atmosphere must stop. To take effective steps to reduce CO_2 emissions, it helps to know which activities produce CO_2 emissions and how much. Direct measurements of CO_2 emissions from specific sources are scarce due to the extra effort and cost required. From the measurement of atmospheric CO_2 concentration alone, we cannot discriminate between the different sources. Therefore, we need to find an indirect method for estimating the contribution of specific sources.

J. Sitte, *Global Warming: Facts, Myths and Remedies*,
https://doi.org/10.1007/978-3-032-20862-0_5

We know that the combustion of fossil fuels (coal, oil, and natural gas) releases large amounts of CO_2. Fortunately, we have records of the quantities of fossil fuels traded and consumed. Because we also know the amount of CO_2 released from burning each type of fossil fuel, we can make useful estimates of the amount of CO_2 released by burning these fuels, where and by whom.

Because the purpose of burning fossil fuels is to obtain energy, in this chapter, we will focus on the amount of CO_2 emitted per unit of energy for each fuel type, rather than per unit mass. This chapter will seek the answer to that question.

5.1 CO_2 Produced by Fossil Fuel Combustion

The fossil fuels we use are coal (solid), oil (liquid), and natural gas (gaseous). The combustion of a unit mass of each fuel releases a different amount of energy and CO_2. For each fuel, the chemical equation for the combustion reaction tells us the mass of CO_2 produced by the combustion. The simplest case is the burning of natural gas, , which mostly consists of methane.

5.1.1 Methane

Methane is the lightest hydrocarbon molecule. The chemical formula of methane is CH_4, which tells us that each methane molecule consists of one carbon atom and four hydrogen atoms. Methane reacts (burns) with oxygen to produce carbon dioxide, water, and energy. The chemical equation for the methane combustion reaction is:

$$CH_4 + 2O_2 \rightarrow CO_2 + 2H_2O + \text{energy} \qquad (5.1)$$

Each methane molecule reacts with two oxygen molecules to produce one carbon dioxide molecule, two water molecules, and some heat. Upon combustion, the mass of one methane molecule begets the mass of one carbon dioxide molecule. Instead of working with the masses of single molecules, which are tiny, it is more convenient to work with *molecular weights.* If you are familiar with molecular weights and moles, you may skip the next section.

5.1.1.1 Molecular Weight

An equation for a chemical reaction like 5.1 tells what happens to individual molecules in the reaction. Therefore, we can calculate the masses of the reaction products from the mass of the reactants. The masses of molecules are tiny when expressed in SI units of mass (kg). Furthermore, we do not burn one methane molecule at a time. In any combustion reaction, we have billions of reactant molecules. The definition of a *molecular weight* simplifies the calculation of the masses involved in chemical reactions. Instead of thinking of one methane molecule undergoing combustion, we will think of so many molecules that we can weigh them on a simple benchtop scale.

The chemical formula of a substance tells us the atoms that comprise a molecule of that substance. Because we know the masses of all the atoms in the periodic table of the elements, it is easy to add the masses of the atoms to get the mass of a molecule. Atoms consist of protons, neutrons, and electrons. The protons and the neutrons make up the nucleus of an atom. The mass of a proton is nearly the same as that of a neutron. The electrons have a much smaller mass than protons; therefore, we can generally ignore their mass in chemical reactions, and we take the mass of an atom to be the sum of the masses of the protons and neutrons in its nucleus.

There is always an integer number of neutrons and protons in an atom; therefore, an atom's mass is a multiple of the proton/neutron mass. The hydrogen nucleus has only a proton, so its atomic weight is the mass of a single proton. The mass of a carbon atom is 12 proton masses because its nucleus has six protons and six neutrons. However, the chemical properties of an element depend on its number of electrons, which is equal to the number of protons. Therefore, we identify the elements by their number of protons and call that number the *atomic number*. In summary, its atomic number (number of protons) identifies an atom, and its mass is a multiple of the proton/neutron mass equal to the sum of the protons and neutrons in its nucleus. The *atomic weight* of the atom is simply the sum of the number of protons and neutrons in its nucleus. If we want the mass of the atom in grams or kilograms, we simply multiply the proton mass by the atomic weight. Similarly, for a molecule, the *molecular weight* is the sum of the atomic weights of its constituent atoms.

For the lighter elements, the number of neutrons in the nucleus of an atom is equal to the number of protons. For the heavier elements, the number of neutrons is somewhat larger than the number of protons. Furthermore, elements have atoms with different numbers of neutrons in their nuclei. These are the isotopes (located in the same place in the periodic table of the elements) of an element, which have the same chemical properties (same number of protons and electrons) but different masses. The most abundant hydrogen atoms have one proton and no neutrons in the nucleus. But there is also the less frequent *Deuterium*, with one neutron in the nucleus. Even rarer is the unstable hydrogen isotope *Tritium*, with two neutrons in the nucleus. Isotopes of elements occur naturally with various relative abundances. Therefore, when we look up the atomic weight of an element, we find that it is not an integer number. The listed atomic weight is the average of the atomic weights of the naturally occurring isotopes of the element, weighted by their natural abundances.

The mass of a proton is very, very small. Consequently, the masses of single atoms are way too small to be measured on a benchtop scale. To get a mass of substance large enough to weigh on a benchtop scale, we need to take a very, very large number of atoms. We work with amounts called moles. A *mole* of substance is the amount of substance that weighs as many grams as its atomic or molecular weight. Therefore, a*mole* of any substance always has the same number of atoms/molecules. The number of atoms or molecules required to form a mole is a very large number, known as Avogadro's Number ($6.02214076\ 10^{23}$). Hydrogen is the lightest element. Its atomic number is one, and the atomic weight of its most abundant isotope is also one. Under normal conditions (room temperature, atmospheric pressure), hydrogen atoms bind in pairs to form hydrogen gas (H_2) molecules. The hydrogen gas molecule has two protons; hence, its *molecular weight* is two, and one mole of H_2 weighs two grams. The carbon dioxide molecule has one carbon atom and two oxygen atoms. The atomic weight of carbon is 12, and the atomic weight of oxygen is 16. Therefore, the molecular weight of CO_2 is 44, and one mole of CO_2 weighs 44 grams.

Armed with the definition of the mole, we can read the reaction Eq. 5.1 as one mole of methane (16 g) and two moles of oxygen (32 g) reacting to produce one mole of carbon dioxide (44 g) and two moles of water (36 g), plus some heat (energy). The heat released by this reaction is 891 kJ.

5.1.1.2 CO_2 Emission

For the combustion of methane, we have that the reaction of 1 mole of methane with 2 moles of oxygen produces 1 mole of CO_2, 2 moles of water, and releases 891 kJ of energy. In other words, burning 16 g of methane produces 44 g of CO_2. That is $44/16 = 2.75$ times the mass of the methane consumed. Because natural gas is mostly methane, burning one tonne of natural gas produces 2.75 tonnes of CO_2. More interesting is to know the amount of CO_2 released when we generate one unit of energy by burning methane. This number allows us to estimate directly the amount of CO_2 pollution from our energy consumption when burning methane. We get the answer by dividing the mass of CO_2 produced by the energy released in the combustion.

$$\frac{44\text{g}}{891\ 10^3 J} = 0.0494\ 10^{-3}\frac{\text{g}}{\text{J}} \approx 50\frac{\mu\text{g}}{\text{J}} \tag{5.2}$$

Ultimately, what matters to us is the energy required to perform a given task, regardless of how we obtain it. We may obtain it from burning gas, kerosene, coal, wood, nuclear reactions, wind, photovoltaic solar, or hydroelectricity. Each source produces different environmental side effects. For atmospheric warming, the fundamental quantity is the CO_2 emitted per unit of energy consumed. This quantity varies across the different types of fossil fuels. It turns out that natural gas is the fossil fuel that emits the least CO_2 per unit of energy generated.

5.1.2 Liquid Fossil Fuels

To extend the calculation to liquid fossil fuels, such as oil and gasoline, we note that these fuels are primarily composed of hydrocarbon compounds, which consist of carbon skeletons with hydrogen atoms attached to the carbon atoms. The combustion of methane, which can be in liquid or gaseous form, produces CO_2 and H_2O. With complete combustion, two oxygen atoms bind to each carbon atom. Therefore, the number of CO_2 molecules produced equals the number of carbon atoms in the hydrocarbon molecule. The combustion of one mole of fuel produces as many moles of CO_2 as carbon atoms are in the fuel molecule. The ratio of the mass of CO_2 to the mass of carbon produced in the combustion of the fuel is

$$\frac{Mass\ of\ CO_2}{Mass\ of\ C} = \frac{12 + 2 \cdot 16}{12} = 3.66 \qquad (5.3)$$

For hydrocarbons, the mass of the carbon in the fuel is nearly the same as the mass of the fuel because the hydrogen atoms are much lighter than the carbon atoms, contributing very little to the total mass. If we include the hydrogen atoms, the fuel mass is slightly greater than the CO_2 mass, and therefore, the CO_2 mass to fuel mass ratio is somewhat less than 3.66. The extreme case is for methane, where the ratio of CO_2 mass to methane mass is

$$\frac{Weight\ of\ CO_2}{Weight\ of\ CH_4} = \frac{12 + 2 \cdot 16}{16} = 2.75 \qquad (5.4)$$

For other fuels, the ratio ranges from 2.75 to 3.7. Most liquid fossil fuels also contain impurities that are either non-combustible or produce other combustion products. For simplicity, it often suffices to apply the rule of thumb that the combustion of 1 tonne of liquid fossil fuel produces around 3 tonnes of CO_2.

Knowing the amount of CO_2 produced per unit of fuel mass allows us to estimate human-caused emissions, as fuel consumption data is readily available. The situation is more complicated for calculating the CO_2 emission per unit of energy supplied for liquid fuels other than methane. The reason is that liquid fossil fuels contain various hydrocarbons that release different amounts of energy per mass when burned. To facilitate this calculation, the International Energy Agency defined the *tonne of oil equivalent* (toe) equal to 41.868 Gigajoules as an energy unit representing the average energy released by burning one tonne of crude oil. Henceforth, we will assume that burning liquid fossil fuel produces, on average, three times its mass of CO_2 and releases one toe of energy. Thus on average the mass of CO_2 released per Joule of energy is

$$\frac{3\,t\ of\ CO_2}{1\ toe} = \frac{3\,t\ of\ CO_2}{41.868\text{GJ}} = 0.0717\ \frac{t\ of\ CO_2}{\text{GJ}} \approx 72\ \frac{\mu\text{g}\ of\ CO_2}{\text{J}} \tag{5.5}$$

This value is close to the widely used value of 73.3 kg/GJ in the literature.

5.1.3 Coal

Coal is a solid that occurs naturally in various forms with different carbon content, from Lignite (brown coal) and Bituminous Coal to Anthracite and Graphite . These forms of natural coal contain water and non-combustible minerals, collectively labelled as *ash*. Anthracite is the highest-grade coal used as fuel, with a carbon content exceeding 85%. Graphite, the pure crystalline form of carbon, is not used as a fuel.

The combustion of coal has several reaction pathways. The simplest of which is the carbon oxidation to carbon dioxide

$$C + O_2 \rightarrow CO_2 + \text{energy} \tag{5.6}$$

The combustion of 1 mole of carbon (12 g) produces 1 mole (44 g) of CO_2. The energy released is 393.5 kJ. Ideally, burning coal would release $44/12 = 3.67$ times its mass in CO_2. Per unit of energy, this makes

$$\frac{44\text{g}}{393.5\ 10^3 J} = 0.112\ 10^{-3}\frac{\text{g}}{\text{J}} \approx 112\frac{\mu\text{GJ}}{\text{g}} \tag{5.7}$$

which is more than double the CO_2 per unit of energy released by natural gas. The CO_2 emission per mass for the combustion of the various types of coal depends on the carbon content of the coal.

5.2 Specific Energy of Fuels

To calculate the amount of CO_2 emitted per unit of energy from burning a specific fossil fuel, we needed to know how much energy it releases per unit mass. We obtained that data from the chemical reaction equation, which tells us the energy released per mole of the reacting substance. Knowing the energy released per unit of fuel volume is also helpful. Therefore, two numbers state how much energy we can obtain from burning a given fuel: the *specific energy* and the *energy density.* The specific energy is the amount of energy released by burning a *unit mass* of the fuel. Energy density is the energy released by burning a *unit volume* of the fuel. The value to use depends on whether we handle the fuel by mass or volume. For example, dispensing fuel by volume is the easiest way at the petrol station. Both the specific energy and energy density have been measured for many substances. Table 5.1 lists the specific energy for various fossil fuels.

Table 5.1 Energy content of various fossil fuels

Fuel	Specific energy MJ/kg
Liquid Natural Gas (LNG)	53.6
Methane	55.6
Gasoline	46.4
Kerosene	43
Coal(anthracite)	26–33
Gas oil/diesel	43.1

To find the amount of CO_2 emission per unit energy for a given fossil fuel, we divide the amount of CO_2 released per unit mass or volume by the specific energy or energy density, respectively.

$$\frac{m_{CO_2}}{energy} = \frac{m_{CO_2}}{m_{fuel}} \cdot \frac{m_{fuel}}{energy} = \frac{m_{CO_2}}{m_{fuel}} \cdot \frac{1}{\frac{energy}{m_{fuel}}} = \frac{m_{CO_2}}{m_{fuel}} \cdot \frac{1}{specific\ energy} \tag{5.8}$$

Where m_{CO_2} stands for the amount of CO_2 produced by burning the fuel, releasing an amount of $energy$. From the carbon content of the fuel, one can calculate the ratio of the mass of carbon dioxide to the mass of fuel. Each mole of carbon in the fuel generates 1 mole of CO_2.

$$\frac{m_{CO_2}}{m_{fuel}} = \frac{m_{CO_2}}{m_C} \cdot \frac{m_C}{m_{fuel}} = \frac{44\,\mathrm{g}}{12\,\mathrm{g}} \cdot \frac{m_C}{m_{fuel}} = 3.67 \cdot \frac{m_C}{m_{fuel}} \tag{5.9}$$

The mass fraction of CO_2 produced by a fuel is 3.67 times the mass fraction of carbon in the fuel.

The value for the CO_2 emission per unit energy for coal calculated in Eq. 5.7 was for the complete combustion of the carbon in the coal. That is not quite how the various types of coal burn. Instead, we can use the measured specific energy for various types of coal and calculate the CO_2 emission per unit of energy using the above equation.

By consulting a chemical reference, we find that the specific energy of anthracite, depending on its quality, ranges from 25 to 33 MJ/kg. For anthracite, the carbon content ranges from 86 to 97% in mass. For Lignite (brown coal), the carbon content is only 60–70%, and the specific energy is only between 10 and 20 MJ/kg due to its high moisture content.

The mass of CO_2 produced per unit of mass depends on how much of the carbon in the coal will burn. We will assume complete combustion of the carbon content. For Lignite with 65% carbon content, we get the ratio

$$\frac{m_{CO_2}}{m_{Lig}} = 3.67\frac{m_C}{m_{Lig}} = 3.67 \cdot 0.65 = 2.38 \qquad (5.10)$$

With an assumed specific energy of 20 MJ/kg, we get

$$\frac{m_{CO_2}}{energy} = \frac{2.38 \cdot m_{Lig}}{energy} = \frac{2.38}{20\ 10^6\,\text{J/kg}} = \frac{0.119}{10^6\,\text{J/kg}} = 119\ 10^{-6}\,\text{g/J} = 119\,\mu\text{g/J} \qquad (5.11)$$

For top-quality anthracite, the same calculation gives

$$\frac{m_{CO_2}}{m_{Ant}} = 3.67\frac{m_C}{m_{Ant}} = 3.67 \cdot 0.97 = 3.56 \qquad (5.12)$$

and

$$\frac{3.56}{33\ 10^6\,\text{J/kg}} = 0.108\ 10^6\,\text{kg/J} = 108\ 10^{-6}\,\text{g/J} = 108\,\mu\text{g/J} \qquad (5.13)$$

These values align with our *ideal* estimate in Eq. 5.7.

Table 5.2 shows the 2023 fossil fuel energy consumption and corresponding average CO_2 emission values for selected countries. The values of CO_2 emission per unit energy in the table are lower than the values for the various fossil fuels that we calculated in this section because some of the energy consumed comes from sources with zero CO_2 emission (solar,

Table 5.2 Calculated average 2023 CO_2 emissions per unit energy from overall fossil fuel energy consumption and corresponding CO_2 emissions for selected countries

Country	Consumption (EJ)	CO_2 emission (Gt)	$\mu g\ CO_2/J$
US	76.00	4.639	61.1
Australia	5.12	0.376	73.4
New Zealand	0.50	0.032	64.0
Germany	8.56	0.572	66.8

hydro, and nuclear). The somewhat different mix of fossil fuels consumed by each country also contributes to variation in the average CO_2 emissions per unit of energy across countries.

5.3 Global CO_2 Emission from Burning Fossil Fuels

Armed with our results for the CO_2 emission per unit of energy from the combustion of fossil fuels, we can estimate the amount of CO_2 released by the 504.83 EJ of energy generated from fossil fuels in 2023. Had this energy come exclusively from burning natural gas, then, using the value of 50 μg/J of CO_2 emission for methane (Eq. 5.2) as the emission for natural gas, this would have produced

$$504.83\ 10^{18}\text{J} \cdot 50\ 10^{-6}\frac{\text{g}}{\text{J}} = 25242\ 10^{12}\text{g} = 25.24\ \text{Gt} \tag{5.14}$$

of CO_2. The EI Statistical Review of World Energy quotes global CO_2 emissions from burning fossil fuels in 2023 as 35.13 Gt. Our estimate is below this amount because the energy content of natural gas, per unit of mass, is around twice that of heavier hydrocarbons such as coal and petrol.

We would have to burn more coal and oil, emitting more CO_2 to get the same energy from burning natural gas. Of the total fossil fuel consumed in 2023, only 28.6% was natural gas. When we take this into account, we get the improved estimate of

$$25.24 \text{ Gt} \times (0.286 + 0.714 * 2) = 25.24 \text{ Gt} * 1.714 = 43.26 \text{ Gt} \tag{5.15}$$

which is just 23% higher than the official, more detailed estimate. Not bad for a back-of-the-envelope calculation! It is quite easy to improve this estimate because the EI Statistical Review of World Energy tabulates global energy consumption for each of the fossil fuels: oil, gas, and coal. Multiplying the energy consumed for each fuel by the values calculated above for the CO_2 emission per Joule, we get the total CO_2 emission for each fuel. Adding those up gives us an improved estimate of 39.7 Gt of CO_2 emitted during 2023. This result is only 13% higher than the estimate from the 2024 EI Statistical Review of World Energy. This result is our reward for the effort of working through the calculations of the amount of CO_2 emitted per unit mass and unit energy from the combustion of fossil fuels. Using published statistical data on fossil fuel consumption, we were able to estimate the world's CO_2 emissions from fossil fuels.

Governments and agencies estimate CO_2 emissions by multiplying the reported masses of the fossil fuel consumed by a chosen factor. That factor is the multiple that gives the mass of CO_2 produced by burning 1 tonne of the corresponding fossil fuel. Above, we estimated an average value of 3 for that factor.

5.4 Fossil Fuels, the Military and Conflict

The global amount of CO_2 emissions from military activities is largely unaccounted for. However, some estimates suggest that approximately 5.5 percent of emissions are caused by the military (Conflict and Environment Observatory, ceobs.org). Therefore, the military sector's fossil fuel consumption should be included in global fossil fuel consumption data. But even this would not include emissions from exploding munitions in armed conflicts.

6

Addicted to Energy

In Chap. 3, we found that the amount of greenhouse gases present in the atmosphere determines the radiation balance and thus sets Earth's global mean surface temperature. Records show that global temperatures are increasing. In Chap. 4, we saw that this warming correlates with the rapid increase of CO_2 released from burning fossil fuels, a process that began with the Industrial Revolution. Today, the wealthiest fraction of the world's population burns large amounts of fossil fuels to generate the energy needed to sustain their comfortable lifestyle.

Almost everything we do requires energy, lots of energy: growing our food and cooking it, refrigerating it, washing our dishes, disposing of our waste, treating our sewage, building, lighting and heating our homes, making our clothes and washing them, pumping the water into our homes and heating it for cleaning, showering and bathing, listening to music, watching TV, browsing the internet, streaming entertainment and playing computer games, talking over the phone, doing work on our computers, driving to work,

J. Sitte, *Global Warming: Facts, Myths and Remedies*,
https://doi.org/10.1007/978-3-032-20862-0_6

travelling to far-flung holiday destinations, manufacturing our cars, appliances and all the myriad of other gadget we must have, sustaining a police force, sustaining an army, building weapons, …and more.

Figure 6.1 shows how quickly the world energy consumption increased in recent history.[1] Moreover, the data on world energy consumption indicate that fossil fuels supply over 80% of the world's energy needs. In this chapter, we will scrutinise the relationship between wealth and energy consumption to determine how we can extricate ourselves from the grip that fossil fuels have on our wealth.

6.1 World Energy Consumption

Determining how much energy humankind consumes and how much comes from fossil fuels is a laborious undertaking. There is no direct "energy meter" from which we can read off the world energy consumption.[2]

To illustrate the problem, let us think about how you could find out how much energy you use to drive to work. On the trip to work, the car consumes energy to overcome road and air friction, heat the engine, accelerate to gain kinetic energy that later dissipates when braking, run the air conditioner, and so on. Determining the energy consumed requires more than just reading the distance travelled from the car's odometer. The energy consumed will depend on how fast you travel, the condition of the road, your tire pressure, how often you accelerate and brake, whether you have the lights on, the number of people travelling with you, and what you carry in the boot. Calculating the contribution of

---

[1] Data before 1965 are from Smil (2017) Data for 1965 onwards are from BP, with Smil's estimates of the traditional biofuels added.

[2] For electric energy, there is. Consumers pay for their electricity according to the readings from their *power meter*.

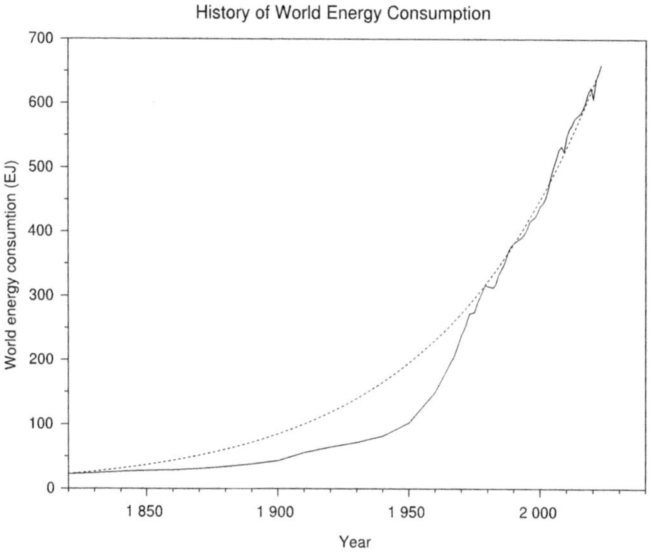

Fig. 6.1 History of world energy consumption since 1800 (solid line). The dotted line shows an exponential growth curve that produces the same increase over the specified period. The energy consumption curve suggests a more complex growth dynamics than simple exponential growth. In particular, the dynamic has changed drastically after World War II. *Source* Statistical review of world energy 2024 (EI 2024) and Smil (2017)

these factors to energy consumption would take a lot of work. A simpler way to estimate energy use is to measure the fuel consumed between visits to the petrol station. Ultimately, burning fuel is what provides the energy for everything. The petrol purchase receipt shows the fuel volume put into the tank. We learned in Chap. 5 how to convert fuel mass or volume to energy. Therewith, we can estimate the energy produced from the fuel consumed, and we have a practical method to estimate world energy consumption from records of fuel traded worldwide.

Many people and institutions laboriously gather data from trade records and generously make it available to the public. In the following analysis, I will utilise data from the annual Statistical Review of World Energy. Until 2024, the review tabulated the *primary* energy consumption for each energy source (fossil fuel, nuclear, hydroelectric, etc.) by country and region. From 2025 onwards, the review tabulates *Total Energy Supply*.[3] The data is all expressed in the SI energy unit of Joule, for easy comparison. British Petroleum (BP) published the *BP Statistical Review of World Energy* from 1952 until 2022, when the *Energy Institute*, a professional society, took over publication responsibility, and became the EI Statistical Review of World Energy(EI 2024) or simply the Statistical Review of World Energy (SRWE).

The definition of *primary* energy is the energy content of the amounts of all commercially traded fuels. However, we also obtain energy from other sources, such as nuclear reactors, hydroelectric power stations, wind turbines, solar panels, geothermal energy, biomass, and more, which do not involve fossil fuels. How can we add the energy from these sources to the fossil fuel energy that we consume to obtain a value for the total energy consumed?

Fortunately, all these non-fossil fuel energy sources produce electricity, and measuring their electric energy *output* is easy. To include non-fossil fuel electric energy in the accounting of energy generated from fuels, until 2024, the Statistical Review of World Energy used a fictitious *input equivalent* for the output. The input equivalent is the amount of fossil fuel a standard thermal power plant would burn to generate the same amount of electricity. To calculate the input equivalent, we must remember that a fossil-fuel electric power plant converts only a fraction of the energy from fossil fuels into

[3]The year of the report is the year of publication, usually in June of that year, and it reports the data for the year before.

electrical energy. The rest is wasted. The SRWE accounts for the loss by using an efficiency factor to calculate the equivalent fuel input from the electricity output. The SRWE utilised an average efficiency factor, updated annually, to reflect the progressive efficiency gains in converting fossil fuels to electricity. For the 2023 data, the efficiency factor was 41.0%. Clearly, this fictitious fossil fuel consumption inflates the total primary energy consumption. Therefore, the Energy Institute, in its 2025 report, switched from reporting *Primary Energy Consumption* to reporting *Total Energy Supply*. Thus, in the 2024 data, energy from *combustible non-fossil fuel* sources, such as nuclear power plants, is counted by their input heat consumption, and energy from *non-combustible* sources by their gross electrical output.

Instead of comparing non-combustible renewable energy sources, such as wind, solar photovoltaic, hydro, ocean, and wave energy, to fossil fuels, they become energy sources in their own right, measured by the gross amount of electricity generated. Still, the problem of converting the output back to input remains. For example, for a wind turbine, the input would be the energy carried by the wind. The chosen solution is to assign a conversion efficiency factor to each source type. For renewables (Hydro, Ocean, Solar PV, Tidal, Wave, and Wind) the EI sets it to 100%. For non-fossil-fuel electricity generation, where the primary energy input is heat, such as nuclear or geothermal energy, which can be difficult to measure, the EI uses efficiencies of 33% for nuclear and concentrating solar, 10% for geothermal, and 33% for the combustion of biomass.

There are no records of the energy obtained from *traditional biofuels*, and therefore, the energy consumption of such fuels is missing in the accounting of world energy consumption. Traditional biofuel is woody matter, dry foliage, and animal dung. Inhabitants of some of the world's rural regions still collect and use significant amounts of traditional biofuel, estimated at around 40 EJ per year (Smil 2017) for cooking and heating. Vaclav Smil uses the notion of *useful energy* to distinguish between primary energy consumption and the energy that does the final useful work because of the loss of energy in the intervening energy transformations. This notion is useful because conversion efficiencies have increased, leading to a faster rise in useful energy consumption than in primary energy consumption. Here, I will not follow this path because of the difficulties in estimating mean conversion efficiencies.

According to the 2024 EI Statistical Review of World Energy(EI 2024), the total energy consumed by the world's population in 2023 was nearly 620 EJ (619.63 EJ), using the *primary energy* method. Out of this total, 18.5% (114.80 EJ) came from non-fossil fuel sources. The remaining 81.5% of the world's energy consumption came from burning fossil fuels. In the 2025 review (data for 2024), using the *Energy Supply* method, that percentage now rises to 87%. The pie chart in Fig. 6.2 starkly illustrates that burning fossil fuels overwhelmingly supplies our energy consumption.

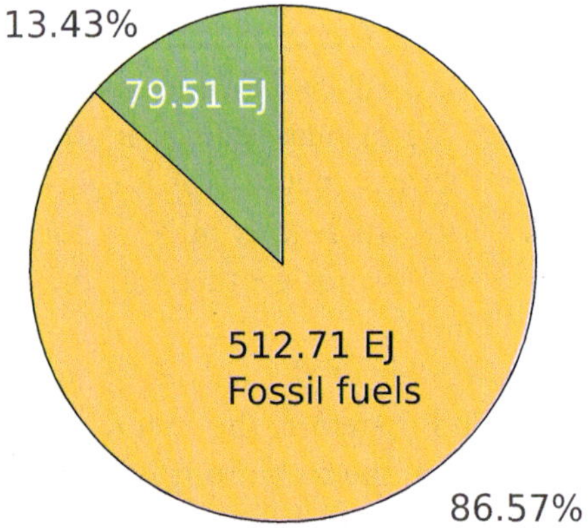

Fig. 6.2 Split of the total 2024 world energy supply of 592.22 EJ into fossil and non-fossil fuel

The world's energy consumption is likely to grow, not only because rich countries increase individual consumption, which they do, but also because the world population continues to grow. Hopefully, poverty reduction will also continue. Poverty reduction means increased energy consumption. In 2022, energy supply increased by 6% from 2020 levels, offsetting the temporary dip in energy consumption in 2020 caused by the COVID-19 pandemic. Table 6.1 shows the growth of primary energy consumption in recent years, and Table 6.2 provides a handy list of important energy consumption related data.

6.1.1 When Energy Is *Consumed, Where Does It Go?*

Think of using your kitchen blender. A force is applied to the blades to cut the veggies into pieces. The blender performs mechanical work by chopping up your veggies. The veggie mix warms up as the blender works. The current through the motor windings' resistance heats the entire motor. Thus, during blending, electrical energy is converted

Table 6.1 Growth of total primary energy consumption from 2015 to 2024

Year	Total (EJ)	Fossil fuels (EJ)	Non-fossil (EJ)	% fossil fuel
2015	550.45	473.38	77.07	86.00
2019	583.90	492.34	91.56	84.32
2020	556.63	462.77	93.85	83.14
2021	595.15	489.66	105.48	82.28
2022	604.04	494.05	109.99	81.79
2023	619.63	504.83	114.80	81.47
2024	592.22	512.71	79.51	86.57

The data are from the corresponding yearly SRWE. Data for 2024 is Total Energy Supply

Table 6.2 Some useful energy numbers

Total world energy consumption (2023)	619.63 EJ	619.63 10^{18}J
World per capita energy consumption (2023)	77.0 GJ	77.0 10^{9} J
Total world fossil fuel energy consumption (2023)	504.83 EJ	504.83 10^{18}J
Average world power consumption (2023)	19.65 TW	19.65 10^{12}W
Total world electricity generation (2023)	107.73 EJ	107.73 10^{18}J
Total world electricity generation from fossil fuel	60.47 EJ	60.47 10^{18}J
Energy equivalent of 1 tonne of crude oil (toe)	41.868 GJ	41.868 10^{9} J
Joule equivalent of 1 kWh (1000 J/s·3600 s	3.60 MJ	3.60 10^{6} J
Energy to heat 1 l of water by 1 degree	4.18 kJ	4.18 10^{3} J

into mechanical energy and heat. Energy conservation is a fundamental tenet of physics. Because energy is conserved, it is still there after we have *used* it. However, we cannot reuse that energy (Second law of thermodynamics).

6.2 Per Capita Energy Consumption

The 2024 world energy supply of nearly 600 EJ per year is a very large number. What that number means is easier to grasp if we express it on an individual scale by dividing it by the world population to get the average per capita consumption. The result tells us that in 2024, every person, on average, consumed 72.56 Giga (billion) Joules (GJ) of energy. Remember, 87% of energ supply came from fossil fuels, which translates to $0.87 \cdot 72.56$ GJ $= 63.13$ GJ of fossil fuel consumption per capita.

How much oil do we have to burn to generate 63.13 GJ? Using the *toe* of 41.868 GJ per tonne, we get

$$\frac{63.13}{41.868\ \mathrm{GJ/t}} = 1.5\,\mathrm{t} \tag{6.1}$$

Thus, in 2024, on average, every person on the planet burned 1.5 tonnes of oil. Since each tonne of burned oil emits about 3 tonnes of CO_2, this means that in 2024, for every person on the planet, we poured 4.5 tonnes of CO_2 into the atmosphere, and we continue to do so.

In the 2015 Paris Climate Accord, the representatives of the countries of the world pledged to curb CO_2 emissions and limit global warming to 1.5° C. Therefore, fossil fuel consumption should have diminished thereafter. Table 6.1 shows this is not the case. Total energy consumption increased from 2015 to 2019 and from 2019 to 2024. One could excuse this by noting that in 2024, there were more people on the planet than in 2015, and in 2024, there were more than in 2019. True, yet on average, each person still consumed more energy in 2024 than in 2015. The per-capita energy supply increased from 70.2 GJ in 2015 to 71.6 GJ in 2019, and then from 68.0 GJ in 2020 to 72.56 GJ in 2024.

Driven by surging per capita energy use and global population growth , the world's total energy supply continues to rise. Does the recent growth of solar and wind energy significantly contribute to reducing fossil fuel consumption? The news is not good. From 2023 to 2024, the fossil fuel energy supply increased by 7.6 EJ, while the non-fossil energy supply increased by only 4.34 EJ. In 2024, the fossil fuel energy supply continued to grow at almost twice the rate of the non-fossil energy supply. The same happened from 2022 to 2023. Despite the growth in non-fossil fuel energy supply, there is still not enough available to meet our growing energy demand, and we are burning more fossil fuels to compensate for the shortfall. Eight years after the Paris Accord, our fossil fuel consumption continued to grow.

6.2.1 What Counts Towards Per Capita Energy Consumption?

The per capita energy consumption not only includes the *direct* individual consumption but also an attribution for the energy consumed *indirectly* by the manufacture of goods, the production of food, and the delivery of services and amenities supplied by the community. Examples of the latter include healthcare, law enforcement, national defense, schooling, and even street lighting. It also counts the energy invested in producing goods for export to other countries. Direct energy consumption mainly comprises household energy consumption for lighting, heating, food preparation, entertainment, computing, etc., and transportation. People have a direct influence over this part of energy consumption. People have much less control over the indirect part of their energy consumption, which is largely determined by the socioeconomic environment.

The direct individual energy consumption of people in affluent countries is around 50 GJ/year. This is about two-thirds of the average world per capita energy consumption. For a person living in the USA, where annual per capita energy consumption in 2023 was a phenomenal 277.3 GJ, direct personal consumption is, on average, only 1/6-th of total consumption. Therefore, for people in affluent countries, changing their direct energy consumption habits will have a modest effect on total energy consumption. If you are one of the wealthy citizens, then cycling to work instead of driving will not save the world.

6.2.2 What Can One Do with 72 GJ?

The per-capita energy consumption of 72.56 GJ in 2024 is still a large number. Let us get a feel for what it means. In Chap. 2, we found that a joule is a relatively small amount of energy. It takes over half of a Mega Joule (568 kJ) to bring the water in a common-size (1.7 l) kettle from room temperature (20° C) to the boil. The world's per-capita energy consumption is over a hundred thousand times that. With that amount of energy, we could bring the water in 128 thousand kettles to a boil. That would be 350 kettles a day, or around one kettle every four minutes. Another way to look at it is to calculate the continuous power consumption that would add up to 72.56 GJ over the course of a year. We get the answer by dividing 72.56 GJ by the number of seconds in the year (see conversion Table 2.1)

$$\frac{72.56\ 10^9\ \mathrm{J}}{3600 \cdot 24 \cdot 365\ \mathrm{s}} = 2301\ \mathrm{W} \tag{6.2}$$

This calculated power can be better understood by comparing it with household appliances. Domestic hot water kettles typically consume 2.2 to 2.4 kW. Thus, the average per capita power consumption is equivalent to having a hot water kettle running nonstop 24/7. Alternatively, a typical oil column room heater consumes as much as your electric kettle. Having a room heater on 24/7 for a year would also add to the 2023 per-capita power consumption.

For comparison, transferring 1 GB of data over the internet consumes an estimated 3.6 MJ (1 kWh). That is 298 Bytes[4] or about two text messages per joule. The average size of a movie download in HD video resolution (720 lines) is 2–4 GB. Thus, a movie download consumes 2–4 kWh, which is equivalent to running your kettle for one hour. Clearly, watching video on demand, or YouTube, is anything but free. At an electricity cost of 20 USD cents per 1 kWh, streaming a movie costs between 40 and 80 USD cents in electricity cost alone. Only a small part of this appears on your electricity bill. Your monthly internet bill, your Netflix subscription, or the ads on YouTube cover the rest of the energy cost. The internet's global energy consumption is significant. According to the IEA, in 2024, Data Centres worldwide consumed 415 TWh (1.5 EJ).

6.3 Energy Consumption and Wealth

Individual energy consumption varies widely. The world average per capita consumption hides who uses more or less than the average. That is the shortcoming of averages. Figure 6.3 shows the per capita consumption for a sample of countries. It shows that for some countries, per capita energy consumption is several times the world average. It also shows that a large portion of the world's population uses far less than the global average. To balance the average, for every person who consumes four times the average, as people in the USA nearly do, there must be twelve people consuming one-third of the average, as people in India do. A person in the USA consumes, on average, the same energy as 12 people in India.

[4] $1GB = 2^{30}$ Bytes.

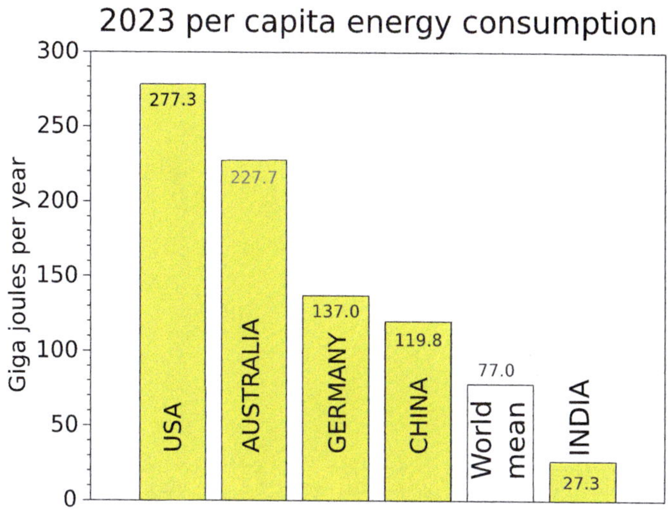

Fig. 6.3 Comparison of the 2023 yearly per capita primary energy consumption of several countries. Data from EI statistical review of world energy (EI 2024)

$$1 \cdot (4 \cdot 77\,\text{GJ}) = 12 \cdot \frac{77\,\text{GJ}}{3} \tag{6.3}$$

A widely used indicator of a country's average wealth is its per capita GDP (Gross Domestic Product). How does per capita GDP relate to per capita energy consumption? The graph in Fig. 6.4 shows the annual per capita energy consumption versus the yearly per capita GDP for various countries for the year 2014.[5] On this graph, we see that the countries form a point cloud that scatters around a rising line. The distribution remains similar regardless of the chosen year.

[5]2014 is the year for which GDP data is available for the largest number of countries.

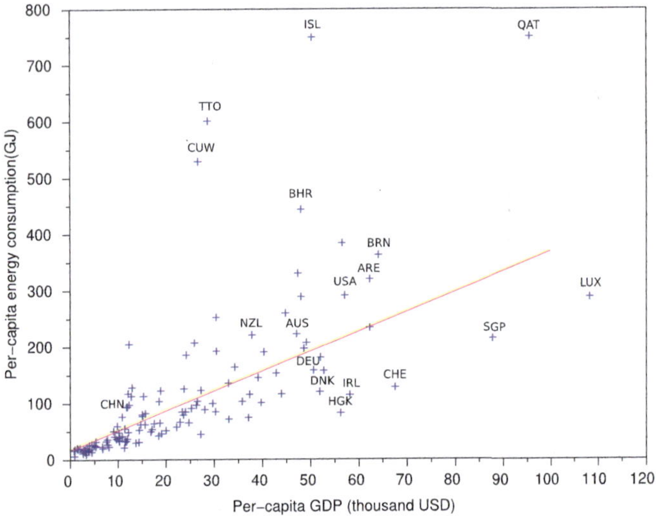

Fig. 6.4 Per capita energy consumption versus per capita GDP in 2014, for most countries in the world. (*Source* World Bank). The linear best fit to the data (solid line) excludes the outliers Qatar (QAT), Iceland (ISL), Trinidad & Tobago (TTO), and Curacao (CUW)

Wealthier countries use more energy. Each country's energy-to-GDP ratio varies due to its unique energy use and efficiency in producing goods and services.

We can also spot some outliers in this distribution. These outliers give some insight into the relationship between energy consumption and wealth. For example, Curacao (CUW) and Trinidad & Tobago (TTO) have medium incomes, but their per capita energy consumption is exceptionally high. The main industry in Curacao is oil refining. In Trinidad & Tobago, it is oil and gas production. Both countries have small populations, so the energy used for export industries is attributed to a small number of people. Iceland and Qatar (QAT) have, by far, the highest per capita energy consumption. Again, they are countries with small populations. Qatar has one of the highest per capita GDPs in the world. Iceland is a wealthy country because its geothermal energy is cheap and plentiful. Iceland derives 85% of its energy from geothermal and hydroelectric resources, which enable Icelanders to thrive in its cold climate. Qatar, with its hot climate, spends much of its energy on cooling rather than heating. However, its wealth comes from oil and gas extraction, which consumes a lot of energy. The citizens of Luxembourg (LUX) are among the wealthiest in the world, despite having per capita energy consumption approximately one-third that of Qatar. The wealth of Luxembourg rests on financial services that consume far less energy than oil and gas extraction.

Countries consume energy to generate their wealth. Some produce more wealth with the same energy. The ratio of per-capita GDP to per-capita energy consumption for a country shows how efficiently its people use energy. For Qatar and Luxembourg, Luxembourg makes three times more income per unit of energy than Qatar! For each country, the ratio also changes somewhat from year to year. This depends on how well the country learns to generate more wealth per unit of energy. Countries have different *learning curves.* In many wealthy industrialized nations, such as Germany, Ireland, and Singapore, per capita energy consumption has declined in recent years while per capita GDP has risen. Countries learn to become more efficient.

The ratio of energy consumption per GDP provides a simple way to calculate a country's CO_2 emissions by using the Yoichi Kaya equation (Kaya and Yokobori 1997).

$$m_{CO_2} = population \times \frac{GDP}{population} \times \frac{Energy}{GDP} \times \frac{m_{CO_2}}{Energy} \qquad (6.4)$$

The ratio of the mass of CO_2 emitted per unit of energy $(\frac{m_{CO_2}}{energy})$ enters as a factor in the expression. To eliminate CO_2 emissions, just reduce this factor to zero. That does not require reducing per capita GDP or energy consumption per GDP. We do not need to reduce the population, average individual wealth, or energy cost of production to eliminate CO_2 emissions. We only need to use emission-free energy sources!

The positive correlation observed between a country's per capita GDP and its per capita energy consumption supports the widespread belief that a comfortable lifestyle is associated with high energy consumption. The thought of reducing energy consumption to lower CO_2 emissions immediately conjures up the fear of losing comfort. Therefore, the idea of stopping global warming by reducing energy consumption meets with widespread rejection, conscious or unconscious, among the population. Remedial action to lower CO_2 emissions should focus on replacing fossil fuels with CO_2-emission-free energy sources, not on reducing overall energy use. In Chap. 7, we will examine candidate energy generation technologies that can replace fossil fuels.

6.4 Income and Happiness

The yearly *World Happiness Report* aims to quantify the average level of satisfaction with life conditions across the world's countries. This provides us with the opportunity to explore the relationship between happiness and income. Using the per capita GDP as published by the World Bank as a proxy for average wealth, Fig. 6.5 shows a graph of the 2025 Happiness score versus Purchase Power Adjusted 2023 per-capita GDP. As expected, the data scatter wildly; yet surprisingly,

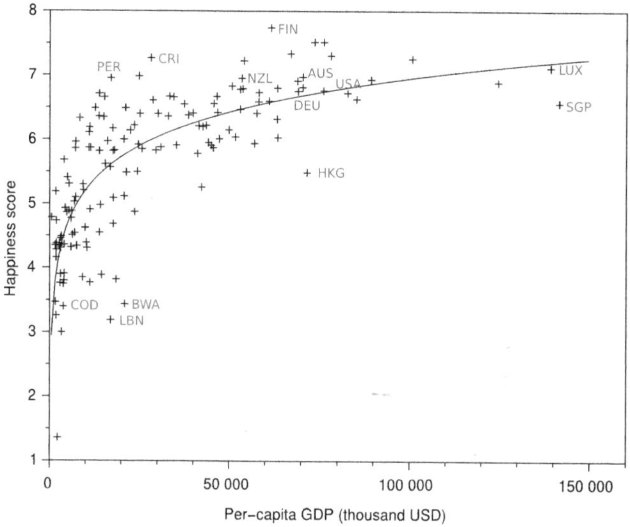

Fig. 6.5 Happiness score against 2023 per capita GDP, purchasing power adjusted, for most countries in the world. (*Source* World Bank). The solid line is a logarithmic curve that best fits the happiness score data (+). (Data from The World Happiness Report 2025)

the Happiness Score appears to depend roughly logarithmically on the per capita GDP. That means happiness increases linearly with the logarithm of the income. In other words, for low-income people, a given increase in income increases their happiness much more than the same income increase would increase the happiness of high-income people.

6.5 The Inequality in Energy Consumption

To see how energy consumption varies across the world's population, I plotted, in Fig. 6.6, the 2014 per capita energy consumption (blue line) against the percentage of the world

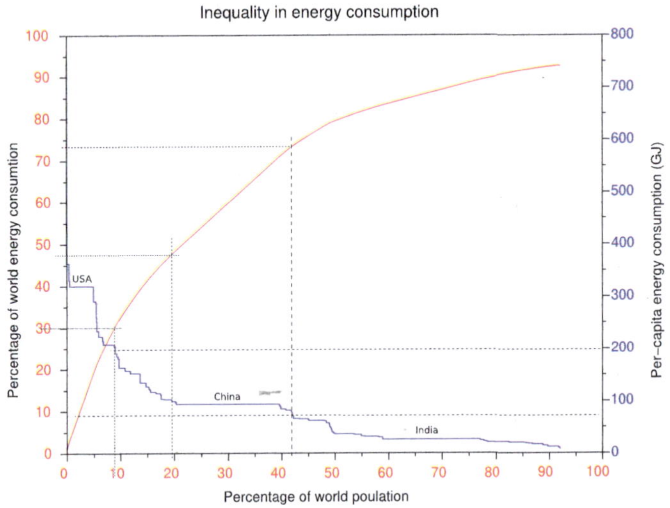

Fig. 6.6 Distribution of energy consumption across the world population. The increasing curve is accumulated consumption, and the decreasing histogram (step curve) is per-capita energy consumption(GJ) in decreasing order

population for 130 countries. The countries are listed in decreasing order of per capita energy consumption. The area under the curve corresponds to total energy consumption. The red line shows the accumulated consumption, up to the percentage of the world population. From that graph, one can read the total energy consumed by any fraction of the highest per-capita energy consumers. For example, 20% of the world's population consumes 48% of the world's energy. The USA, with only 5% of the world's population, consumes nearly 20% of the world's energy.

It also shows that in 2014, although only 42% of the world's population (three billion people) lived in countries with above-average per capita energy consumption (74 GJ), their joint energy consumption accounted for 73% of world consumption. That means that more than half (58%) of the population shared the remaining quarter (27%) of the energy. Furthermore, 9% of the population had per capita energy consumption exceeding 200 GJ, accounting for 30% of global energy consumption.

The differences in per capita energy consumption reflect disparities in national wealth. Thus, the wealthiest nations consumed the most energy and, consequently, have contributed overwhelmingly to global warming through their CO_2 emissions into the atmosphere.

6.6 Energy Consumption and International Trade

As we discussed earlier, the total per capita consumption includes the energy that a country uses to produce goods for export. This energy should be counted as consumed by the people in the country that buys the goods. However, the revenue from exported goods increases the wealth of the exporting country (GDP). The exporting country spends part of this wealth on purchasing goods from other countries. These imported goods also required energy to produce. In the long run, a country cannot spend more on imports than it earns from exports. Every country must aim to balance its trade. However, that balance is monetary. It is difficult to equate the monetary value of goods with the energy used in their production. We may assume that, over time, the energy the exporter consumes for the importer's use will largely be offset by the energy the importer expends for their exports. In this way, the exported energy consumption eventually *comes home*. Also, by shifting energy-intensive manufacturing to developing countries, industrialized countries reduce their energy consumption and, consequently, a portion of their greenhouse gas emissions.

Table 6.3 Comparison of 2019 imports/exports for China and the USA

Item	China	USA
Energy consumption (EJ)	141.7	94.7
GDP (billion USD)	14,280	21,430
Exports (billion USD)	2,499.5	1,644.3
Imports (billion USD)	2,069	2,567.5
Population (million)	1,398	328.3
Energy consumption per capita (GJ)	98.8	287.6
GDP per capita (USD)	10,216.6	65,279.5
Exports per capita (USD)	1,787.9	5,008.5
Imports per capita (USD)	1,480	7,820.6

China, as the world's manufacturer, exports its wares to the rest of the world. Thus, when estimating China's energy consumption, we attribute the energy used to produce its export goods to China, although it should be attributed to the importing countries. In turn, China buys raw materials and food from the rest of the world. Counting the energy spent by the country that produces these supplies for China as energy consumed by that country compensates somewhat for the energy embodied in the wares imported from China that were not accounted for in the country's energy consumption.

From the data in Table 6.3 we read that China's 2019 energy consumption, at 141.7 EJ, was 1.5 times that of the USA's, at 94.65 EJ. Furthermore, China's total exports in 2019, at USD 2,499.48 billion, were also 1.52 times those of the USA, at USD 1644.28 billion. Unfortunately, we do not know how much energy is attributable to the production of export goods in either China or the United States. When it comes to the GDP generated per unit of energy consumed, China's is less than half that of the USA's. The USA is more *efficient* in converting energy into income.

In 2019, Chinese people, with a per capita GDP of 15.7% of that of the people in the USA, had much less money available for buying imported goods. China's total imports approach those of the USA, largely due to its large population. Likewise, China's large population enables it to produce 1.5 times the export value of the USA, while per capita energy consumption is one-third that of the USA.

When it comes to CO_2 emissions, in 2019 China obtained 85% of its energy from fossil fuels. China emitted CO_2 for the production of export goods on behalf of importing countries. Importers, in turn, needed to export goods and also emitted CO_2 to pay for these imports, effectively emitting on behalf of their own client countries. Therefore, while these emissions may not offset each other entirely, the net impact of international trade on CO_2 emission attribution can, in a first approximation, be considered negligible. In 2022, China was the country with the highest CO_2 emission. It emitted 10,523 Gt of CO_2. With a population of 1.43 billion people, this makes 7.36 tonnes per capita. The USA emitted 4.7011 Gt of CO_2 in 2022. With a population of 0.34 billion people, this amounts to 13.9 tonnes per capita, almost twice China's per capita emissions.

6.7 Energy Consumption Targets

Energy is the most important resource supporting comfortable human living conditions. Achieving a comfortable living standard for all of humanity is a declared ethical goal formulated in the 2030 Agenda for Sustainable Development, adopted by the United Nations in 2015. It declares the eradication of poverty as its Number One goal. This goal clearly implies higher energy consumption among the world's poorest people. The "Energy for Growth Hub" (Energy for Growth Hub 2023) estimates that 3 billion people suffer from *energy poverty*. It sets a per capita target of 1000 kWh/year (36 GJ/year) of available electricity that would allow a **dignified and productive** life for everyone.

As long as energy production primarily comes from burning fossil fuels, which is most likely the case until 2030, achieving the minimum energy consumption target means increasing CO_2 emissions. From Fig. 6.6, we see that lifting the yearly energy consumption to a minimum of 36 GJ means an increase in energy consumption for about 40% of the world's population. Clearly, achieving this without increasing the world's energy consumption requires a large reduction in energy consumption by the world's largest energy consumers. To put this into context, Bangladesh's per capita energy consumption was 9.6 GJ, while Singapore's was 609.5 GJ.

6.8 Energy Waste

As long as the world consumes large amounts of fossil fuels, wasting energy results in unnecessary CO_2 emissions. Once we have abundant, CO_2 emission-free energy sources, wasting energy becomes less important. Until such time, increasing energy efficiency is a significant way to reduce CO_2 emissions. As we shall find in 9, replacing internal combustion engines with electric motors to do mechanical work drastically reduces energy waste. Energy waste is widespread in our technical civilisation, and consequently, there are many ways to reduce energy consumption without substantial loss of quality of life. Internet streaming and cloud computing are other examples. As mentioned in Sect. 6.2.2 above, the internet is power-hungry. Too many of us are unaware of this fact, which allows a small number of large transnational corporations to generate large profits. The *convenience* that the *Cloud* provides comes at a significant price in CO_2

emissions. Although the energy per bit of internet transmission is in decline, the growth in the amount of data transmitted far outweighs this decline.

7

Alternatives to Fossil Fuels

The analysis in the previous chapters strongly suggests that to stop global warming now, we must immediately renounce all fossil fuel consumption. For most people, this is an unpalatable solution because it would mean a drastic drop in living standards.

A better proposition is to **replace** fossil fuel energy with energy sources that do not emit CO_2 and therefore would not restrict our energy consumption. Fortunately, humankind has the know-how and the capacity to make that substitution. Various potential CO_2- emission-free energy sources are receiving attention in public discussion. In this chapter, we investigate each of these and assess their technical maturity, economic viability, environmental impact, and their ability to supply the required energy. Let us start with the much-talked-about CO_2 emission-free fuel: Hydrogen.

J. Sitte, *Global Warming: Facts, Myths and Remedies*,
https://doi.org/10.1007/978-3-032-20862-0_7

7.1 The Hydrogen Economy

The idea of a *Hydrogen Economy*, that would use hydrogen as a universal fuel, is widely publicised and has the backing of a strong industrial lobby. The fact that nurtures the dream of a Hydrogen Economy is that hydrogen combustion releases energy without producing carbon dioxide. *Burning* hydrogen consumes oxygen, which is abundant in the atmosphere, releases energy and produces water vapour. The emission of water vapour is deemed harmless to the environment. Natural gas burners and internal combustion engines can run on hydrogen with only minor modifications, and *fuel cells* convert hydrogen chemical energy directly to electricity. Thus, hydrogen appears to be the ultimate *clean* fuel. Proponents of the Hydrogen Economy like to add that hydrogen is the most common element on Earth and in the universe. Despite this statement being true, hydrogen is not a readily available fuel. Furthermore, the use of hydrogen, as we shall see below, comes with some significant *inconveniences.*

7.1.1 Hydrogen, Some Basic Facts

Burning hydrogen means letting it react with oxygen to produce water and energy, according to the following chemical equation:

$$2H_2 + O_2 \rightarrow 2H_2O + \text{energy} \tag{7.1}$$

The reaction of 2 mol (4 g) of hydrogen with 1 mol (32 g) of oxygen to form 2 mol (36 g) of water releases 572 kJ of energy. Thus, the specific energy for hydrogen is $572\,\text{kJ}/4\,\text{g} = 143\,\text{kJ}/\text{g} = 143\,\text{MJ}/\text{kg}$, which is nearly three times the $55,69\,\text{MJ}/\text{kg}$ specific energy of natural gas (which is mostly methane). For the same mass, hydrogen carries

Table 7.1 Hydrogen as an energy carrier compared to methane

Fuel	Heat of combustion (kJ/mol)	Specific energy (kJ/g)	Energy density (kJ/l)
Hydrogen	286	143.0	12.77
Methane	891	55.7	39.78

almost three times as much energy as methane. That sounds very good until we realise that the same mass of hydrogen occupies a much larger volume than the same mass of methane because hydrogen is much lighter. At standard conditions of temperature and pressure (STP),[1] both hydrogen and methane are in a gaseous state, and a mole of hydrogen and a mole of methane have the same volume (22.4 l). However, as Table 7.1 shows, a mole of hydrogen carries only about one-third of the energy of a mole of methane. Therefore, measured per volume (energy density), hydrogen carries less than one-third of the energy of methane. Transporting a given amount of energy with hydrogen requires three times as much space as with methane.

Cooling down methane and hydrogen at atmospheric pressure until they condense into liquid greatly reduces their volume. Transoceanic refrigerated shipping of liquefied natural gas (LNG) is standard practice. Doing the same with hydrogen is far more costly because hydrogen has a much lower boiling point than methane. Liquefaction of hydrogen can consume over 30% of its energy content, depending on the liquefaction process and machinery used. Hydrogen liquefaction is also a well-established process, as is the safe handling and storage of hydrogen, though it is quite costly.

[1] Standard conditions of temperature and pressure STP is 0°C and atmospheric pressure at sea level.

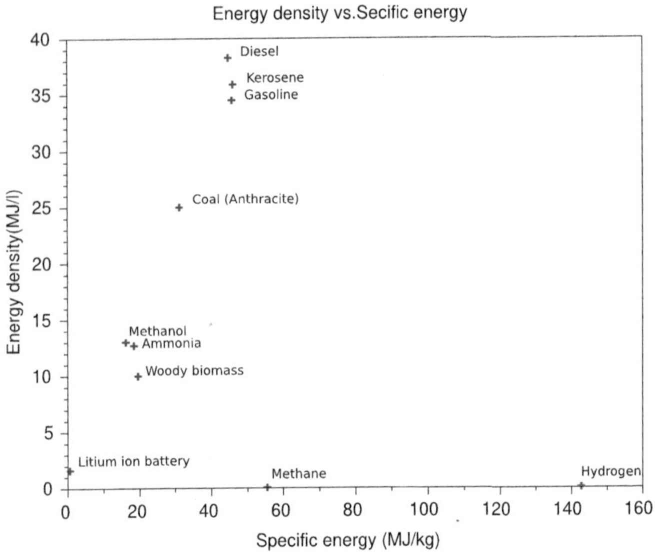

Fig. 7.1 Energy density versus specific energy for various fuels, at ambient temperature

For any fuel, the specific energy (energy per unit mass) and the energy density (energy per unit volume) are the properties that matter for the storage and transport of the fuel. Figure 7.1 compares these characteristics for various fuels. It shows that hydrogen has the highest specific energy but the lowest energy density, which makes it the most difficult to store and transport.

7.1.2 Where Do We Find Hydrogen?

The idea of a *Hydrogen Economy* has been around since the early 1970s. Why do we not yet live in a Hydrogen Economy?

The first and foremost reason is that, despite being abundant as an element, there are no large deposits of hydrogen gas ready to extract and burn. Fossil fuels are found almost ready to burn. Hydrogen is not, although it might be. It was as recent as 2012 when geologists confirmed that the gas emanating from a well in Mali was 98% hydrogen. Now, geologists suspect that water seeping deep into the Earth's mantle oxidises hot iron deposits, thereby releasing hydrogen. The released hydrogen may accumulate under impermeable layers. Whether exploiting such deposits would be profitable is still unknown.

For the time being, we know of very little free molecular hydrogen (H_2) available. All hydrogen is bound in other molecules, such as water and hydrocarbons. To obtain molecular hydrogen, we need to remove hydrogen atoms from hydrogen-containing molecules. And that costs energy!

Thus, if we want large quantities of hydrogen, we have to chemically *extract* it from hydrogen-containing substances. For example, we can split water into oxygen gas and hydrogen gas by electrolysis. Splitting water by electrolysis takes about 260 kJ/mol. We can recover that energy by letting the hydrogen *burn* with oxygen to become bound in water again. The fact that we can recover the energy spent to generate hydrogen gas is what makes hydrogen a *store* of energy. Obviously, because hydrogen is a store of energy and not a source, producing hydrogen with energy from fossil fuels does not reduce our CO_2 emissions.

7.1.3 Example: Driving on Hydrogen

Looking at what it would take to fuel cars with hydrogen illustrates the *inconveniences* of hydrogen fuel. An average ICE car can travel 500 km on a full fifty-litre tank of fuel. Let's calculate how much hydrogen we need to put in the tank to drive 500 km in a hydrogen-powered vehicle.

We could adapt the motor of a conventional ICE car to run on hydrogen, but an electric vehicle (EV) uses energy more efficiently. Electric motors are over 90% efficient in converting electric energy into mechanical energy, compared to the 20–30% efficiency of an ICE. To generate the required electricity to power the EV, we can use a *fuel cell.* Fuel cells convert the chemical energy stored in hydrogen gas into electricity.

Manufacturers claim that an average battery EV (BEV) consumes around 16 kWh per 100 km. 16 KWh is 57.6 MJ. We need more energy if we use a fuel cell instead of a battery. Fuel cells are about 60% efficient, and we would need 57.6 MJ/0.6 = 96 MJ to travel 100 km in the fuel cell EV (FCEV). For 500 km, we need 480 MJ, which equals the energy content of 3.4 kg of hydrogen. The FCEV's tank should hold at least 3.4 kg of hydrogen.

The first hurdle to putting 3.5 kg of hydrogen in a tank is that one kg of hydrogen gas at STP occupies 11,200 l (11.2 m^3). To drive the 500 km, we need 38 m^3 of hydrogen gas! Storing and moving such large volumes of gas is impractical. To carry the 38 m^3 would require a trailer with the width of the car (1.8 m), 2 m high, and 10.6 m long—hardly a practical proposition. So, let us try to compress the hydrogen to fit into a typical 50l tank of a petrol car. To fit into a 50 l tank, we need to compress the hydrogen to 50/38,000 = 1/760 of its original volume, which takes a pressure of 760 atmospheres (remember the law of ideal gases). For comparison, this pressure is 25 times the test pressure of 30 atmospheres (bars) for a propane gas steel bottle. The hydrogen tank has to be much, much stronger. We can build high-pressure tanks. Tanks for FCEVs made out of carbon fibre composite that can store around 4 kg of hydrogen and withstand a pressure of 700 atmospheres weigh around 100 kg. Although such containers are heavier, bigger, and more expensive than the 50 l petrol tank, they still fit into an FCEV.

An alternative way to reduce the volume of a gas is to cool it until it condenses into a liquid. For natural gas, this is a proven method. By cooling methane at atmospheric pressure to −162 ° C, it becomes a liquid that occupies about 2 litres per kilogram. To do the same with hydrogen gas, one must cool it to 20.28 K (−252.87° C). Once cooled, it can be stored for some time at atmospheric pressure in thermally well-insulated containers.

The liquefaction temperature for hydrogen is about 100° C lower than the liquefaction temperature of methane and is only 20° C above absolute zero. Liquefying hydrogen consumes a lot more energy than compression. In practice, it takes more than 40 MJ/kg, or nearly a third of its specific energy, to liquefy hydrogen. Furthermore, there are inevitable evaporation losses from liquid hydrogen stored at atmospheric pressure in highly thermally isolated containers. Thus, storing liquid hydrogen in an FCEV looks highly impractical. Once we have resolved how to store hydrogen in the tank, we still need to think about how to transport it from the production site to the tank. First, we need a refilling station capable of filling hundreds of vehicles. We can avoid transporting the gas to the refilling station by installing electrolysers that produce the hydrogen on-site and dispense it. The refilling station will not need a large storage tank, but it will need a compressor. The electrolyser needs to be powered by renewable energy (solar or wind), and so must the compressor. Both electrolysers and hydrogen compressors have efficiencies of around 70%. Thus, there are non-recoverable energy losses throughout the process of storing and recovering the energy. From the energy needed to run the electrolyser, we only store 70% in the hydrogen. The compressor leaves us with another 70% of that. Finally, the fuel cell returns only 60% of the energy put in. Overall, we recovered $0.7 \cdot 0.7 \cdot 0.6 = 0.3$ times the energy put in. To store the 480 MJ of energy needed to drive 500 km, we would need to use approximately three times as much solar electricity (480/0.3 = 1600 MJ). Last but not least, each step costs money.

7.2 Insisting on Hydrogen

A Hydrogen Economy is attractive because it would not contribute to global warming, provided all hydrogen production is CO_2- emission-free. Achieving this requires an abundant supply of CO_2 emission-free energy. For example, producing hydrogen through water electrolysis requires electricity from renewable sources, such as solar, wind, or geothermal power stations. If we already have renewable energy, why not use it directly rather than convert it to hydrogen and waste energy in the process? Renewable electricity sources, such as solar and wind, are intermittent. We need some way to store renewable electricity. Hydrogen can serve as a means to store energy for use when the sun isn't shining or the wind isn't blowing.

As found above, powering an electric vehicle with hydrogen wastes half ($0.7 \cdot 0.7 = 0.49$) of the energy stored in the hydrogen gas before it even reaches the fuel cell. Storing renewable electricity as hydrogen means losing more than half of that energy. There are better ways of storing electric energy, such as batteries and pumped hydroelectric dams, which can recover a larger fraction of the stored energy. Pumped hydro can recover over 80% of stored energy, while batteries can recover over 90%. By comparison, from hydrogen, even in the best cases, we recover only about 40% of its specific energy (143 MJ/kg). For reference, in 2020, a high-capacity Li-ion battery only stored roughly 0.9 MJ/kg; however, we could recover almost all of it.

Batteries still offer only limited energy storage capacity, making them impractical for storing large amounts for extended periods. In turn, pumped hydroelectric dams can store large amounts of energy almost indefinitely, but they require a power grid to distribute electricity, which limits the availability of electricity to the next grid connection point. Hydrogen stores large amounts of energy for extended periods and can be carried by mobile energy consumers, almost like conventional fuels. But, low energy recovery and complex handling only make sense where batteries are insufficient or the grid limits mobility, such as heavy machinery and aviation. Hydrogen energy storage makes economic sense only for a few cases where other renewable electricity storage options fail to meet requirements.

Worldwide, large-scale industrial processes consume approximately 100 million tonnes (Mt) of hydrogen annually. Over half of it goes into producing ammonia to fertilise the crops that feed the world population. Petroleum refining consumes another quarter. The remainder goes into the production of steel and other chemicals, such as methanol. As we shall see below, a good use of hydrogen made with renewable electricity would be to replace the large amounts of 'dirty' hydrogen currently used in many industrial processes.

Advocates of the hydrogen economy aim for *clean* hydrogen to cost less than USD 2.00 per kilogram to compete with fossil fuels. With hydrogen's specific energy at 143 MJ/kg, to meet this target, its production cost would have to be:

$$\frac{2\ \text{USD}}{143\text{MJ}} = 0.014\frac{\text{USD}}{\text{MJ}} \tag{7.2}$$

or 0.05 USD/kWh.

Compared with fossil fuels, a barrel of crude oil weighs 136 kg. At a cost of USD 80/barrel, it makes 80/0.136 = 588.2 USD/tonne. The toe is 41.868 GJ/tonne, so the cost is then 588.2 USD/41868 MJ = 0.014 USD/MJ (0.033 USD/kWh). Given the similar conversion efficiencies of hydrogen and crude oil, at 2 UDS/kg, hydrogen would be comparable to fossil fuels at USD 80 per barrel. However, that target remains elusive.

The prevailing method of producing hydrogen is by steam reformation of natural gas or coal. Steam reformation of natural gas, which is mostly methane, consists of heating the gas with steam (water) at high pressure in the presence of a catalyst to produce hydrogen and carbon monoxide:

$$CH_4 + H_2O + 206\,kJ \rightarrow CO + 3H_2 \tag{7.3}$$

In the water-gas shift reaction, the CO from the steam reformation reaction further reacts with steam to produce CO_2 and more hydrogen.

$$CO + H_2O \rightarrow CO_2 + H_2 \tag{7.4}$$

The reactions are endothermic and thus need a supply of energy (heat).

If the reaction's heat comes from burning fossil fuel, the CO_2 from combustion adds to the CO_2 from steam reformation. As a result, coal or natural gas steam reformation emits CO_2, even with renewable energy like solar electricity providing the reaction heat.

Another method is to split water (H_2O) into oxygen and hydrogen by electrolysis. Electrolysis of water produces no CO_2. Producing hydrogen through water electrolysis powered by renewable electricity stores renewable energy without CO_2 emissions. Methane pyrolysis offers another CO_2-free method to produce hydrogen. Heating methane with a catalyst separates hydrogen from carbon, resulting in hydrogen gas and solid carbon. Again, the process is only emission-free if powered by renewable energy. Methane pyrolysis is not yet mature for large-scale hydrogen production (Sanchez Bastardo 2021).

The hydrogen debate now uses colors to indicate the origin of the hydrogen. Green, blue, and grey are some of the colours associated with hydrogen. Hydrogen made by methane or coal steam reformation is *grey* when all the CO_2 emitted in its production is released into the atmosphere. Hydrogen is *blue* when the CO_2 emitted was captured. Hydrogen is *green* when no CO_2 was emitted in its production. Currently, the only way to produce green hydrogen is through the electrolysis of water with renewable electricity.

Although grey hydrogen is unacceptable as a clean fuel, blue hydrogen is the darling of coal and oil producers for mitigating global warming because it offers them a way to extend the life of their industry. They want to use their fossil fuels to produce hydrogen and promise to make it clean by capturing all the CO_2 emitted in the production.

Hydrogen's low overall energy storage efficiency makes the hydrogen economy a poor idea. It can only compete with renewable electricity in cases where electricity is not viable, such as long-haul aviation and heavy mobile machinery.

7.2.1 Hydrogen Derivative Fuels

There are other simple chemicals that burn without adding CO_2 to the atmosphere, and these can be made from hydrogen gas. When synthesised from green hydrogen, they become potential green fuels. One example is ammonia (NH_3); another is methanol (CH_3OH). Ammonia burns to form nitrogen gas and water. Methanol, although it burns to form water and CO_2, releases only the CO_2 used in its production, making its use carbon neutral. These substances are attractive green fuels because they are much easier to store and transport than hydrogen. However, this advantage comes at the cost of extra energy losses during their synthesis from green hydrogen.

7.2.1.1 Ammonia

Ammonia , synthesised from green hydrogen and nitrogen , is an attractive alternative energy carrier to hydrogen. Ammonia is already an industrial chemical produced in high volume worldwide, second only to sulfuric acid, and the infrastructure for ammonia storage and transport already exists. At room temperature, ammonia is a gas that liquefies at −33.3° C. Thus, liquefying ammonia by cooling or compression for transport and storage is much easier than liquefying hydrogen. Burning ammonia releases energy without CO_2 emission, producing nitrogen and water. Nitrogen makes up 80% of the Earth's atmosphere and is not a greenhouse gas.

Now let us look at how much *net* energy we can store in ammonia produced from green hydrogen. The combustion reaction of ammonia is

$$4NH_3 + 3O_2 \rightarrow 2N_2 + 6H_2O + 1267.20\,\text{kJ} \qquad (7.5)$$

Per mole the combustion releases $(1276.2/4)\,\text{kJ/mol} = 316.8\,\text{kJ/mol}$, compared to 286 kJ/mol for hydrogen. As moles of any gas occupy the same volume at the same temperature and pressure, *the energy density* of ammonia is thus slightly higher than for hydrogen

$$\frac{316.8\,\text{kJ}}{286\,\text{kJ}} = 1.108 \qquad (7.6)$$

The *specific energy* though, at

$$\frac{316.8\frac{\text{kJ}}{\text{mol}}}{17\frac{\text{g}}{\text{mol}}} = 18.64\frac{\text{kJ}}{\text{g}} = 18.65\frac{\text{MJ}}{\text{kg}} \qquad (7.7)$$

is only about one-tenth that of hydrogen and less than half (0.43) of that of gasoil.

When liquefied, values improve. Liquid ammonia's density at its boiling point is 681.9 g/l, giving an energy density of:

$$\frac{681.9\,\text{g/l}}{17\,\text{g/mol}} \cdot 316.8\,\text{kJ/mol} = 12.71\,\text{kJ/l} \qquad (7.8)$$

While this is only about one-third of the energy density of gasoil, it is still high enough for practical transport and storage.

The reaction for the synthesis of ammonia from nitrogen and hydrogen over an iron catalyst (Haber process) is

$$N_2 + 3H_2 + 91.8\,\text{kJ} \rightarrow 2NH_3 \qquad (7.9)$$

Synthesizing ammonia from hydrogen and nitrogen requires 91.8 kJ/mol, plus 260 kJ/mol for hydrogen production by electrolysis, for a total of 351.8 kJ/mol. Burning ammonia recovers 316.8 kJ/mol. This excludes energy overheads from process inefficiencies.

Fertiliser production accounts for over 80% of global ammonia production. Producing ammonia from solar hydrogen enables small-scale fertiliser production. Small plants can combine PV solar panels, an electrolyser, and an ammonia production reactor to make fertiliser locally. The challenge is to make this process cheaper than using fossil fuels for ammonia. One advantage of small local production is saving on high shipping costs from large fossil-fuel-based ammonia plants. Excess solar energy from large solar farms can also be used for ammonia synthesis.

7.2.1.2 Methanol

Burning methanol produces water, CO_2, and energy. Methanol serves as an energy store and a *green* fuel if the CO_2 produced by burning methanol does no more than restore the CO_2 taken out of the atmosphere during the methanol production. With due care, this is possible, as we shall see below.

Methanol is used to produce a wide range of industrial chemicals, and therefore, the infrastructure for handling and storing methanol is already in place. Methanol is a liquid at room temperature, boiling at 64.7° C, and is easy to store and transport. Industrial scale methanol production is by synthesis from *syngas* (synthetic gas), a mixture of CO and H_2, with hydrogen typically from steam reformation of fossil fuels or biomass. Note that CO_2 will be emitted during methanol synthesis unless the hydrogen in the syngas is generated from a green source.

The combustion reaction of methanol is

$$2CH_3OH + 3O_2 \rightarrow 4H_2O + 2CO_2 + 1366.91\,\mathrm{kJ} \tag{7.10}$$

Per mole the combustion releases $1366.91/2\,\mathrm{kJ/mol} = 683.46\,\mathrm{kJ/mol}$, compared to 286 kJ/mol for hydrogen.

The specific energy

$$\frac{683.46\frac{\mathrm{kJ}}{\mathrm{mol}}}{32\frac{\mathrm{g}}{\mathrm{mol}}} = 21.36\frac{\mathrm{kJ}}{\mathrm{g}} = 21.36\frac{\mathrm{MJ}}{\mathrm{kg}} \tag{7.11}$$

is somewhat higher than ammonia's, but only about half as high as gasoil's and about one-sixth as high as hydrogen's. Methanol's density at room temperature is 792 g/l, resulting in an energy density of

$$\frac{792\frac{\text{g}}{\text{l}}}{32\frac{\text{g}}{\text{mol}}} \cdot 683.46\frac{\text{kJ}}{\text{mol}} = 16.92\frac{\text{MJ}}{\text{l}} \tag{7.12}$$

Although slightly higher than ammonia's, methanol's energy density is lower than gasoil's; transporting the same energy requires three times the volume of methanol as gasoil.

Methanol is conventionally made from *syngas* (synthetic gas), a mixture of CO and H_2. The synthesis reaction is

$$CO + 2H_2 \rightarrow CH_3OH + 91\,\text{kJ} \tag{7.13}$$

Note that the reaction is exothermic.

An alternative synthesis reaction

$$CO_2 + 3H_2 \rightarrow CH_3OH + 2H_2 + 50\,\text{kJ} \tag{7.14}$$

that used CO_2 directly (Bowker 2019) is more costly but has the advantage of using CO_2 that can come from the capture from CO_2 released in other industrial processes or from direct capture of CO_2 from the atmosphere.

Using hydrogen, even from renewables, for energy storage incurs significant energy losses. However, green hydrogen can help mitigate global warming by replacing grey hydrogen in the production of chemicals such as ammonia and methanol.

7.3 Nuclear Fission Power

Commercial nuclear power generation has been with us since 1957. What is usually meant by nuclear power generation is the production of thermal energy by splitting (fission) the nuclei of heavy atoms. In a nuclear power plant, the thermal energy released in its nuclear *reactor* heats steam to drive the on-site electric generators. The nuclear power plant generates electricity without emission of CO_2. In this sense, nuclear energy is *clean* energy.

The construction and operation of nuclear fission reactors is a mature industry that nevertheless has serious handicaps.

1. Although the nuclear power industry claims that modern nuclear power plants are safe, severe accidents have undermined public confidence.
2. Nuclear power plants need a supply of fissile material (fuel),[2] which is of limited availability.
3. Nuclear power plants need large amounts of cooling water, which restricts their location.
4. The atomic nuclei produced by fission are unstable and remain radioactive for thousands of years. The long-term, safe storage of radioactive waste is an unsolved problem.

According to a study by Abbott (2011), it may not be possible to scale up nuclear power generation to meet current global energy consumption for various reasons, including a lack of fissile material (uranium ore). Although there are enough proven deposits of fissile material, the low concentrations of many deposits may make their extraction uneconomical.

7.4 Nuclear Fusion Power

The fusion of light-atom nuclei also releases large amounts of energy. The energy of the Sun comes from the fusion of hydrogen nuclei to produce helium nuclei. The advantages of fusion are that, first, hydrogen is abundant, and

[2] Matter whose atomic nuclei can be split to release energy (nuclear fission).

second, the fusion product (helium) is stable; hence, there is no radioactive waste. The fusion of hydrogen nuclei is thus a *clean* source of energy. The reason we do not have nuclear fusion power plants is that nuclear fusion is difficult to control because the extreme pressures required to bring nuclei close enough to fuse are hard to achieve.

For over 60 years, all attempts to build a fusion reactor that sustainably outputs more energy than is put in have failed. Undeterred, the proponents of the technology continue to claim that success is only twenty years away. The only successful human-made realisation of nuclear fusion is the hydrogen bomb. The hydrogen bomb releases the fusion energy in a short, uncontrolled burst (which is not helpful for generating electricity!). Even so, it is worthwhile to continue trying with the hope that one day, it may become a practical way to generate abundant clean energy. There is considerable research and development activity focused on building nuclear fusion reactors. There seem to be encouraging *proof-of-concept* results. Still, the time needed for commercial deployment and installation of the required energy generation capacity is likely to exceed by far the time we have available to replace fossil fuels.

7.5 CO_2 Capture

Carbon capture and sequestration (CCS) refers to the process of capturing the CO_2 produced in the combustion of fossil fuels and storing it away so that it cannot escape into the atmosphere. Understandably, the fossil fuel industry strongly advocates this as a solution to global warming. CO_2 has been injected into geological strata for several decades, mainly to improve oil recovery. To be viable as an emission reduction procedure, CO_2 must be collected directly at the source of large-scale emissions. Despite being tried and promoted for quite some time, the technology has so far only succeeded in capturing around 0.1% of total CO_2 emissions. Many of the CCS installations in operation barely capture about half of the emitted CO_2.

If Humanity fails to drastically reduce fossil fuel consumption by the end of the next decade, then CCS is the last resort to slow, or even reverse, global warming. To stop global warming with CCS alone would require the capture of all of the 35Gt of CO_2 emitted per year from burning fossil fuels. Reverting global warming requires capturing significantly more than that.

For large-scale CO_2 capture, it is necessary to concentrate the emissions sources to a manageable number. The electrification of domestic, medium-sized industries, agriculture, and transport (excluding air and sea transport) could achieve this. Fossil fuel-fed power plants capturing the CO_2 at the source could meet the demand for electric power. Large industrial plants, such as cement factories, would also have to capture their CO_2 at the source. Whether this transformation of the energy sector is achievable in a short time (a decade) is unknown. Relying on the fast development and deployment of CCS would be an enormous gamble, to put it mildly.

7.6 Solar Energy

Solar energy stands out among alternative energy sources for its potential to provide a permanent, CO_2- free energy supply capable of sustaining a comfortable lifestyle for all humankind. The amount of energy the Earth receives from solar irradiation is thousands of times larger than humankind's current energy consumption. From the Earth's energy balance, which we studied in Sect. 3.4, we found that from the average solar energy flux of $340.4\,W/m^2$, incident on the top of the Earth's atmosphere, 186.3 W/m^2 (54.8%) reaches the Earth's surface. When multiplied by the Earth's surface, that number gives a total power of

$$186.3 \frac{W}{m^2} \cdot 510.06\ 10^{12} m^2 = 95.02\ 10^{15} W \qquad (7.15)$$

With 31.536 million seconds in a year, this makes a total yearly energy of

$$95.02\ 10^{15} W \cdot 31.536\ 10^6 s = 2.997\ 10^{24} J \approx 3\ 10^6 EJ \qquad (7.16)$$

This calculation indicates that solar energy can provide 4842 times the world's energy consumption for 2023. Therefore, the sun offers us an abundance of energy that we can harness in various ways.

Indeed, the sun is the source of energy for all life on our planet. Until the discovery of nuclear energy, human civilisation had relied exclusively on energy from the sun, captured by photosynthesis in plant biomass . Once humans discovered how to make fire, they could release solar energy stored in biomass by burning it. Although burning wood and straw releases CO_2, it does not increase the CO_2 content of the atmosphere because it only returns to the atmosphere the CO_2 that was extracted earlier from the atmosphere when building the biomass by photosynthesis. That changed when we discovered that we can burn fossilised biomass (coal and petroleum), deposited over millions of years. The CO_2 released from burning these fossil fuels was captured too long ago and too slowly to compensate for the sudden release of CO_2 from burning large amounts of fossilised biomass (fossil fuels). The CO_2 from burning fossil fuels accumulates in the atmosphere with the consequence that we already know: the warming of the climate.

Fortunately, we have found other ways of capturing solar energy that do not release CO_2. We can capture solar energy by letting the sun heat matter, such as water, as in thermal solar panels, or by generating steam in solar thermal power plants. The sun warms the atmosphere, and the resulting air currents (wind) can drive wind turbines for electricity generation. Solar energy evaporates water that falls as rain. We can capture the rainwater in elevated dams and use it to drive turbines to generate electricity. Lately, we have invented Photovoltaic (PV) solar cells to convert solar radiation directly into electricity. These are all proven technologies that are available to meet our energy demand. We can stop harmful CO_2 emissions because we have other ways to obtain energy besides burning fossilised or fresh biomass.

Hydroelectric dams, wind turbines, and photovoltaic solar cells convert incoming solar radiation directly into electricity. Almost every energy end-use can be electrified directly or indirectly. Indirect conversion involves using electricity to produce an intermediate energy carrier, such as hydrogen, which can later be converted back into electricity when needed. The challenge is to scale up hydroelectric, wind, and solar electricity generation to replace the fossil fuels that still account for 81.5% of the world's energy consumption. Such replacement is no longer a technical problem but rather an economic and production capacity problem.

The main obstacle for direct solar electricity generation is its intermittency. PV solar cells only deliver electricity when the sun shines on them. But when the sun shines, we need energy the least, while when it does not, we need it the most. Although the wind may blow when the sun does not shine, extended periods of calm are common. The obvious solution to intermittency is energy storage that bridges the gaps when the sun does not shine or the wind does not blow. Hydroelectric dams overcome intermittency because they inherently store energy.

First and foremost, there are batteries for storing electric energy. Next, there is chemical energy storage by producing intermediate fuels, such as hydrogen and ammonia, that can be burnt to generate electricity when needed. Both hydrogen and ammonia burn without releasing CO_2. Finally, there is mechanical storage, where electric motors lift a mass to store mechanical potential energy, which, when released to fall, can drive electric generators. Pumped hydroelectricity, , which is pumping water into an elevated reservoir, is one way to do this. When needed, the water is left to run back down through turbines to generate electricity. A smaller-scale version involves lifting concrete blocks weighing a tonne or more to a height of 20–30 m, then releasing them to drive a generator as they come back down.

Charging batteries when solar electricity is plentiful is convenient for local or mobile energy storage. Battery costs have fallen so much that solar cells paired with battery storage now rival the cost of power from fossil fuel plants. The same is true for wind, though wind turbines are less suited for small-scale distributed electricity generation than solar cells. Batteries require relatively scarce resources and have shorter lifetimes than solar cells. Replacement needs generate waste if batteries are not recycled. The existing technology of Grid Forming Inverters enables the integration of traditional synchronous generators and renewable electricity sources (inverter sources) into the grid without compromising power grid stability. It is only a matter of scaling up the GFM's capacity to large utility-scale PV and wind-battery storage sources.

7.7 Geothermal Energy

Geothermal energy already supplies large amounts of emission-free thermal energy for heating and electricity generation in places such as Iceland, where volcanism brings the heat from deep in the Earth to the surface. Geothermal energy is available only in a limited number of locations. Therefore, it is a valuable contribution to the clean energy supply, but it will not become a mainstream energy source. Unless recent advances in deep drilling—which could enable access to thermal energy almost anywhere—are successfully developed and demonstrated at a production scale. The technology for this method is still in its infancy and has not yet been proven at a commercial level.

7.8 Biomass

Plants capture solar energy by photosynthesis all the time, and we can exploit it for energy generation. Burning wood (biomass) for heating and cooking is humankind's oldest way of generating energy. The specific energy of dry woody biomass ranges from 16 to 20 MJ/kg, , which is less than half that of fossil fuels. Burning biomass for energy releases CO_2 into the atmosphere, just as burning fossil fuels does. Biomass is a net-zero CO_2 emission energy source only over its complete life cycle. The reason is that the CO_2 released to the atmosphere by burning cancels out with the CO_2 absorbed from the atmosphere while growing. What renders biomass, or biofuels for that matter, unsuitable as a *clean* energy replacement for fossil fuels is the low efficiency of plants in capturing solar energy. Optimistic estimates for biomass yield from forest plantations range from 7.5 tonnes per hectare per year to 20 tonnes/(ha y) (Berndes et al. 2003). Let us calculate the rate of energy production (power) per square metre with the optimistic values of 20 tonnes/(ha y) for forest dry mass yield and a specific energy of 20 MJ/kg.

$$20\frac{tonnes}{ha\ y}\cdot 20\frac{\mathrm{MJ}}{\mathrm{kg}} = \frac{20{,}000\,\mathrm{kg}}{10{,}000\,\mathrm{m}^2\cdot 31.536\ 10^6\mathrm{s}}\cdot 20\frac{\mathrm{MJ}}{\mathrm{kg}} =$$
$$1.27\frac{\mathrm{J}}{\mathrm{m}^2\,\mathrm{s}} = 1.27\frac{\mathrm{W}}{\mathrm{m}^2} \qquad (7.17)$$

From this, we still need to subtract the energy spent on cultivating, harvesting, and chopping the wood for burning or for producing biofuels, such as methanol. If we want to convert biomass to electric energy, we incur an additional 60% loss at the power plant. Compare this with an estimated $34.72\,\mathrm{W/m^2}$ electricity generated by a PV solar panel (see Sect. 8.2.1), and the absurdity of using biomass to substitute fossil fuels becomes obvious. Generating enough biomass to meet energy demands would require vast areas of land. The land used for biomass energy production competes with land needed for food production or simply capturing CO_2 (without later release) from the atmosphere.

7.9 End-User Energy from Fossil Fuels

Now that we have an estimate of the amount of usable electricity we can generate from PV solar panels, we also want to determine the required PV solar power capacity to replace *all* of our fossil fuel consumption. Total world energy consumption during 2023 was 619.63 Exajoules. Of these, 81.5%, that is 504.83 Exajoules (EJ), were obtained from burning fossil fuels.

As pointed out in Sect. 6.1, the fossil fuel energy consumption is calculated from the amount of fossil fuel consumed, which includes the energy lost in the transformation to useful end energy. Therefore, fossil fuel energy consumption exceeds the energy delivered to the end user. For example, a fossil fuel-driven electric power plant has an efficiency of around 40%, which means that only 40% of the energy from the fossil fuel burned in the plant is converted into electric energy. PV solar panels instead deliver electric energy directly to the end user. To replace the fossil-fuel power plant with solar electricity, we only need to generate an electric output equivalent to 40% of the power plant's energy input.

Converting fossil fuel energy into mechanical work is much less efficient than converting electric energy. The internal combustion engines that power vehicles and machines are only 20–30% efficient. Electric motors convert over 80% of the electricity they consume into mechanical work. For some applications, it is easy to estimate the amount of electric energy required to replace fossil fuel consumption, for example, by switching from petrol-fuelled cars to electric cars (see Sect. 9.2). In other cases, it isn't as easy to estimate the electric energy needed to replace fossil fuels entirely.

To illustrate this complexity, consider estimating the electric energy required for a Trans-Pacific flight with a CO_2 emission-free passenger aircraft. The propulsion technology is still uncertain; for example, if hydrogen is used, we must determine hydrogen consumption per kilometre and the energy needed for green hydrogen production. This falls beyond our immediate scope but highlights the intricacies of estimating total replacement needs. Despite these hurdles, let us now make a rough estimate of the total useful energy required to replace all fossil fuel consumption.

We use fossil fuels for many purposes: generating electricity in power plants, producing mechanical energy in engines and turbines (in ground, sea, and air transport), heating rooms, and providing heat for industrial processes. Data about the distribution of energy consumption across different uses is scarce. The IEA Key World Energy Statistics 2021

(IEA 2021) contains data on the distribution of total energy consumption across sectors for 19 IEA member countries.

For instance, IEA data show that in 2019, the transport sector accounted for 35% of global energy consumption, with passenger cars accounting for 21%. In 2019, the passenger car fleet was almost entirely petrol-fuelled. We know that an electric vehicle uses less than one-fifth the energy of an ICE car. In this case, we can say that for 21% of the world's energy consumption, we need to replace it with only 20% of the fossil fuel energy consumed. For the remaining 14% of the transport sector (air, sea, rail, and trucking), no data is available. Part of rail transport is already electrified. Sea freight uses predominantly diesel engines. Ammonia-fueled combustion engines could replace them. The overall efficiency may not be as high as that of electric cars, but it could be comparable to the efficiency of an electric power plant.

Turning to heating, the IEA reports that space heating accounts for approximately 20% of global energy consumption. Fossil fuels still provide 64% of the heating energy. Fossil fuel heaters achieve an efficiency above 90%; therefore, nearly all the input energy must be replaced by electric energy. While electric resistance heating converts 100% of electric energy into heat, heat pumps can transfer up to three times as much thermal energy as the electric energy they consume. Good insulation can also make a significant contribution to energy savings. Therefore, electrifying heating could still substantially reduce overall energy use despite the high efficiency of fossil-fuel heaters.

In 2023, world electricity generation was 107.73 EJ (29,924.8 TWh). With a power plant efficiency of 40%, it would have required the consumption of 107.73 EJ/0.4 = 269,3 EJ of fossil fuel, which is equivalent to 43% of the world's primary energy consumption (620 EJ). In other words, for replacing 43% of the world's fossil fuel consumption (269,3 EJ) with solar electricity, we only need to generate 107.73 EJ, or 40% of the fossil fuel energy used for electric power generation.

From these examples, we observe a wide range across sectors in the percentage of fossil fuel input that can be replaced by solar electricity to fully offset fossil fuel consumption. It is around 20% for transportation, 40% for electricity generation from fossil fuels, and around 90% for space heating. If, in the absence of detailed data, we assume 50% as the average for the electric energy needed to fully replace fossil fuel consumption, we should not be too far off the mark. Henceforth, we will assume that we need to generate 50% of our fossil fuel consumption with solar electricity to replace it entirely.

Since electricity generation capacity is typically expressed in power terms, we should also express fossil fuel energy consumption in terms of average continuous power. We now calculate the average 2023 fossil fuel power consumption P_{FF} by dividing the 2023 fossil fuel energy consumption by the number of seconds in a year.

$$P_{FF} = \frac{504.83\ 10^{18}\text{J}}{365 \cdot 24 \cdot 3600\ \text{s}} = 15.920\ 10^{12}\text{W} = 15.92\ TW \tag{7.18}$$

Using our estimate of average 50% of conversion efficiency, the effective power P_e obtained from the consumed fossil fuel power P_{FF} is

$$P_e = 0.5 \cdot P_{FF} = 0.5 \cdot 15.920\ 10^{12}\text{W} = 7.96\ 10^{12}\text{W} \tag{7.19}$$

To generate this amount of power with PV solar panels, we need an installed solar peak power of

$$P_p = \frac{P_e}{\epsilon} = \frac{7.96\ 10^{12}\text{W}}{0.193} = 41.24\ TWp \tag{7.20}$$

It is worth noting that some of the solar electricity will be further converted to mechanical energy. Thus, there will be some loss in this conversion, and the amount of solar electric power has to be somewhat higher than the 41.24 TWp calculated above. However, electric motors are highly efficient; therefore, for the time being, we will ignore the energy loss incurred in this conversion.

7.10 Zero Emission Life-Cycle

For whatever zero-carbon technology we choose to replace fossil fuels with for electricity generation, we must not forget the greenhouse gas emissions throughout the entire life cycle of the selected electricity generator. This includes initial manufacture, operation, and disposal. A technology is truly zero-carbon if its net lifecycle emissions are zero. This can only be achieved if the energy used in production comes from zero-carbon sources. For PV solar cells, this means the cells must be manufactured and recycled entirely with electricity from zero-carbon sources, for example, by using solar electricity to manufacture and recycle them. This is a chicken-and-egg situation. Ideally, we need to *bootstrap* solar cell production from scratch! This is clearly possible, and we have already advanced on this path.

7.11 The Bottom Line

Investing in inferior alternatives to fossil fuels due to existing vested interests or ideological bias diverts funds from the rapid scaling up of truly effective alternatives. Having found that solar energy is abundant and that the main technical obstacles have been overcome or are about to be overcome raises the next question: can we afford to transition our energy use from fossil fuels to solar energy, and can we do it fast enough to prevent substantial further global warming? We will investigate the feasibility in the next chapter.

8

Solar Power

In Sect. 7.6, we found that the sun provides us with much more energy than we currently consume and that we can expect to consume in the future. Our exploration, in Chap. 7, of the candidate energy sources for replacing fossil fuels made it clear that PV solar electricity, together with solar thermal energy and wind, are the only technologies mature enough to be deployed at the scale needed to meet the world's energy consumption. In this chapter, we will take a closer look at solar energy technology and what we need to do to halt global warming by replacing the world's current fossil fuel consumption with solar photovoltaic and wind electricity, as well as solar thermal energy. The two ways of capturing solar energy, heat capture and direct conversion to electricity, differ in their efficiency, that is, the fraction of incoming solar irradiation they capture. We shall investigate the distinctive advantages and disadvantages of each method.

J. Sitte, *Global Warming: Facts, Myths and Remedies*,
https://doi.org/10.1007/978-3-032-20862-0_8

8.1 Thermal Solar Energy

Thermal solar energy capture is already in widespread use for household and industrial hot water. Due to its high efficiency, typically from 70 to 80%, thermal solar energy capture needs far less roof area than solar PV panels to capture the same amount of energy. A wide variety of technologies are used to capture and store thermal solar energy. Solar heating techniques range from recirculating warm air from roof spaces to utilizing the thermal capacity of walls to store heat, or even heating stone pits with warm air during the day for release at night. In temperate and cold climates, heating accounts for a large part of people's total energy consumption. In the European Union in 2015, half of the total energy consumed was for heating and cooling (HRE4 2017). Solar thermal capture can contribute a significant fraction of energy consumption without requiring conversion to electricity.

8.2 Photovoltaic Solar Electricity

Photovoltaic solar energy capture uses *solar cells* to convert sunlight directly to electricity. With electricity, we can power almost anything, and we can even convert it back to heat. Over the last century, we have also built extensive infrastructure for generating, transmitting, and using electric energy. Because photovoltaic solar electricity generation is already an established, mature technology, it is the obvious candidate for replacing fossil fuels for most energy uses. To know whether we can scale up PV solar energy generation to cover world energy consumption, we must answer the following questions:

- Can current photovoltaic solar panels produce the needed energy?
- Do we have space for the panels?
- Can we pay for the fabrication and installation of the panels?
- Do we have the resources to fabricate the panels?
- Are there any harmful side effects?

8.2.1 Energy Generation Capacity of Solar Panels

The average solar flux that reaches the Earth's surface is 186.3 W (see Table 3.1). How much of that flux can we capture and convert to electric energy?

At any chosen location on Earth, the solar radiation flux varies significantly throughout the day, across the seasons, and with the weather. At sea level, on a clear day at noon, a surface facing the sun receives a peak daytime solar irradiation of around 1 kW/m^2, and none during the night! The best commercially mass-produced photovoltaic (PV) solar panels convert slightly more than 20% of the incoming solar flux into electric energy.

The conversion efficiency of PV solar cells is rather low in comparison with the 70–80% conversion efficiency of thermal (hot water) solar panels. The reason for this *low* efficiency is inherent in the technology. The semiconductor material that converts solar irradiation into electricity in crystalline photovoltaic solar cells only absorbs light at some discrete wavelengths. In this sense, solar cells are similar to greenhouse gases, which absorb radiation at a few specific spectral lines. The difference from greenhouse gases is that, with solar cells, we want to capture the incoming solar radiation in the visible range, where the intensity is highest. Because the cells can only convert a narrow range of wavelengths into electric current, they are transparent to much of the solar radiation. By applying various tricks, it is possible to increase conversion efficiency somewhat. For example, using two or more layers of photovoltaic material, each absorbing light at a different wavelength. Tandem solar cells, featuring a perovskite layer that absorbs blue light and a silicon layer that absorbs red light, have achieved laboratory efficiencies of approximately 32%. By placing a reflective layer on the back of the cell, forcing the light to pass through the cell twice, the amount of light absorbed is also increased somewhat. Increasing solar cell efficiency is a topic of intense research and development. However, due to the nature of the photovoltaic process, it is highly unlikely that efficiencies comparable to those of thermal capture will ever be attained.

For our calculations, we will use a panel conversion efficiency of 18%, thereby allowing for some internal loss in the photo-current collection circuitry. Thus, from a 1 m^2 solar panel mounted to face the sun (see Fig. 8.1) with clear skies at noon, we may generate 180 W of *peak* electric power.

As the sun moves across the sky from sunrise to sunset, the angle of incidence of the solar flux onto the surface of a fixed solar panel goes from grazing incidence, that is, incidence at a low angle nearly parallel to the surface, to perpendicular incidence and back to grazing incidence. The more the direction to the sun deviates from perpendicular to the panel surface, the smaller the area of the solar flux intercepted by the panel. The area of solar flux intercepted varies with the cosine of the angle between the direction to the sun and the direction perpendicular to the panel.[1]

[1] Tilting the panels to keep them perpendicular to the sun will maintain maximum solar irradiation for one panel. However, for an array of panels, tilting them will produce shadows on the neighbouring panel that offset the gain.

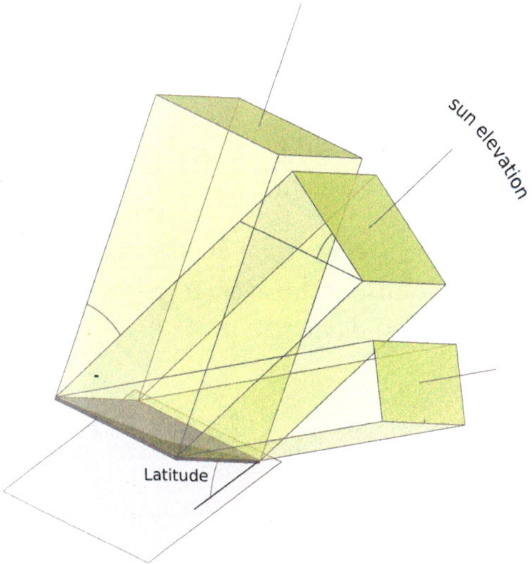

Fig. 8.1 Area of sunlight intercepted by a panel for various elevations of the sun from morning to midday

Over the 12 h of daylight, the panel intercepts about two-thirds (0.64) of the full solar flux. The number of hours of full sunshine per day also varies from place to place on Earth and is far less than 12 h per day. For example, Rome receives approximately 2400 h of sunshine per year, which translates to an average of 6.5 h of sunshine per day, or 0.54 of the average 12 h of daylight. Assuming that the hours of sunshine are distributed evenly across the average 12 h of daylight, then the average incident flux over the daylight time is

$$1\ \frac{\mathrm{kW}}{\mathrm{m}^2} \cdot 0.64 \cdot 0.54 = 345.6;\ \frac{\mathrm{W}}{\mathrm{m}^2} \tag{8.1}$$

The solar panel conversion efficiency of 18% leaves us with an average power of

$$0.18 \cdot 345.6; \frac{\mathrm{kW}}{\mathrm{m}^2} = 62.208; \frac{\mathrm{kW}}{\mathrm{m}^2} \tag{8.2}$$

To calculate the average total energy generated over 12 h of daylight, we multiply the average power by the number of seconds in 12 h of daylight

$$62.208 \cdot \frac{\mathrm{W}}{\mathrm{m}^2} \cdot 3600\frac{\mathrm{s}}{\mathrm{h}} \cdot 12\,\mathrm{h} = 2.687\frac{\mathrm{MJ}}{\mathrm{m}^2} \tag{8.3}$$

Solar panels produce electricity even on cloudy days. Therefore, we will raise the estimate of the average obtainable energy to 3 $\mathrm{MJ/m^2}$ per day.

We can come to the same result by using the average solar flux at the ground (186.3 W, from the NASA energy budget; see Table 3.1) and multiplying it by the number of seconds in a day, instead of accounting for diurnal, seasonal, and geographical variations

$$186.3\frac{\mathrm{W}}{\mathrm{m}^2} \cdot (3600 \cdot 24)\mathrm{s} = 16.096\frac{\mathrm{MJ}}{\mathrm{m}^2\mathrm{day}} \tag{8.4}$$

which when multiplied by the 18% solar panel efficiency gives us again around 3 $\mathrm{MJ/m^2}$ per day (2897 MJ to be exact).

To compare solar panel electricity output regardless of weather and geographical location, manufacturers rate their solar panels by their peak output, which is when the sun irradiates the panel perpendicularly on a clear day, at 1 $\mathrm{kW/m^2}$. A one square metre panel with an efficiency of 18%, receiving 1 kW, will generate 180 W and will therefore be rated as 180 Wp (*peak power*). For all further estimates, we will assume that a panel rated at 180 Wp generates 3 MJ of electric energy per day and occupies an area of 1 m^2.

How much power does the end user actually receive? Spreading the 3 MJ evenly over 24 h makes an average *effective power* of

$$3\,\frac{\text{MJ}}{\text{day}} = \frac{3 \cdot 10^6\,\text{J}}{3600 \cdot 24\,\text{s}} = 34.72\ \text{W} \qquad (8.5)$$

Thus, a panel rated at 180 Wp supplies enough energy to continuously power a 34.72 W load. However, the output of the panel varies throughout the day, with peak power occurring at noon under clear skies and zero output during the night. To maintain a continuous supply of 34.72 W, excess energy generated when output exceeds 34.72 W must be stored, and stored energy must be drawn upon when output falls below this mean value.

The ratio ϵ of effective output to nominal output

$$\epsilon = \frac{34.72\ \text{W}}{180\ \text{Wp}} = 0.193 \qquad (8.6)$$

is a useful number for converting peak power P_p into average effective continuous power P_e

$$P_e = \epsilon\, P_p \qquad (8.7)$$

We will use this value as a world average in our calculations, remembering that the effective power output depends on where on Earth we install the panel and that the value in Eq. (8.6) is an estimated average for Rome.

8.3 PV Solar Energy Land Requirement

Solar panels take up space. How much land area would we need to install the 40 TWp solar panel capacity (see Sect. 9.9) required to replace the average 2023 world fossil fuel power consumption?

Solar panel installations typically require some free space around each panel for access and a supporting structure. Therefore, the space utilisation will not be 100%. Let's make a reasonable guess of 80% space utilization again. This means that a 1 m^2 solar panel will occupy on average 1.25 m^2 and instead of getting 180 Wp/m^2 we only get $180 \cdot 0.8 = 144$ Wp/m^2. The area required to generate 40 TWp is

$$\frac{40 \cdot 10^{12}\ \mathrm{Wp}}{144\ \frac{\mathrm{Wp}}{\mathrm{m}^2}} = 0.278 \cdot 10^{12}\mathrm{m}^2 = 2.78 \cdot 10^5\mathrm{km}^2 \tag{8.8}$$

We would need to set aside around 278,000 square kilometres to install the solar panels required to replace the world's fossil fuel energy consumption. This area is slightly less than Italy's area, about 3.6% of Australia's area, and about 0.19% of Earth's land surface. From this rough estimate, it is clear that the land area required to supply the whole world's energy demand with PV solar panels is modest. A fraction of the Australian outback or the Sahara desert would suffice.

Because solar panels can also be installed on rooftops, the need for land dedicated exclusively to solar panels would be significantly reduced. Furthermore, *agrovoltaics* is being trialled and is found useful. Agrivoltaics (as reported in Petroni 2023) consists of growing crops under elevated solar panels.

What we have not yet estimated is the space required to store generated solar energy for consumption when the sun is not shining. We can safely assume that it will be a fraction of the space occupied by the solar panels. In many cases, the storage may be in batteries placed directly underneath the panels.

In 2020, a high-performance Cobalt Li-ion battery had an energy density of 1.6 MJ/l. To store ten days of energy generated by a one-square-metre panel, we need a volume of 30MJ/1.6MJ/l = 18.75 l. We can easily store ten days of generated solar energy in a battery the size of two standard 10l buckets underneath the panel. Or, if we prefer pumped solar energy storage in a dam, the potential energy per m^2 of the water in a 10m deep reservoir at 300 m above a turbine stores about 30 MJ/m^2, which is 10 times the daily energy output per m^2 of a solar panel. Because hydroelectric turbines have an efficiency of 90%, almost all the potential energy stored in the water can be converted into electric energy.

8.4 Solar Panel Cost

What would be the cost of installing the required 40 TWp solar panel capacity? Data from the U.S. National Renewable Energy Laboratory (NREL) (Ramasamy et al. 2023) indicate that in 2023, the average price per watt peak across domestic, commercial, and utility installations was around USD 1.50. According to this estimate, the installation of 40 TWp would cost

$$\begin{aligned} 40\ 10^{12}\ \text{Wp}\ \times 1.5\ \frac{\text{USD}}{\text{Wp}} &= 60\ 10^{12}\ \text{USD} \\ &= 60\ \text{trillion USD} \end{aligned} \tag{8.9}$$

For comparison, the world GDP for 2023 was USD 105.4 10^{12} (105.4 trillion). The cost of installing 40 TWp of solar panels would be about half (57%) of the world's 2023 GDP. This cost is more than we can afford, and besides that, we do not have the capacity to install that many panels in a year. If we spread the installation over 10 years, that would be approximately 6% of GDP per year. If every country on Earth would dedicate 6% of its GDP to the installation of PV solar cells, in 10 years, every person on Earth would have 70 GJ (91% of the 2023 average per capita energy consumption[2]) electric energy and the world CO_2 emissions would be zero!

Note also that demanding a percentage of GDP investment in solar PV energy is a fair approach . From the data in Fig. 6.4, we know that a country's energy consumption correlates positively with GDP. Therefore, a larger GDP means higher energy consumption and greater CO_2 emissions.

To put that 6% of GDP into perspective, according to the World Health Organisation, global health spending in 2022 was 9.9% of global GDP. This indicates that spending 6% of GDP over 10 years is a substantial financial commitment, yet it remains feasible. And, of course, we must ask, what would be the cost of letting global warming go on?

Neither must we forget that we already spend large amounts of money on fossil fuels. As solar energy replaces fossil fuels, that expenditure would gradually decrease to nil. So, how much does the world spend on fossil fuels every year? The problem with finding such a number is that each fossil fuel—coal, oil, and gas—has its own price that varies widely due to fluctuations in supply and demand. Then there are government taxes and excises that differ from country to country. Lastly, numerous government subsidies artificially reduce prices. Estimates put these subsidies at half a trillion (500 billion) USD per year worldwide. Clearly, estimating the world's annual expenditure on fossil fuels is a daunting accounting exercise.

[2]By 2033 the world population is projected to be 8.8 billion people.

Nevertheless, to get an idea of the order of magnitude, let us make a simple estimate. The average spot crude oil price in 2023 was around USD 82 per barrel (thus, we need to calculate in barrels). Taking the *barrel of oil energy equivalent* of 6.119 GJ (EI 2024), we can obtain the number of barrels required to meet all of the 2023 fossil fuel energy consumption of 504.83 $\approx$ 505 EJ.

$$\frac{504.83\ 10^{18}\mathrm{J}}{6.119\ 10^{9}\mathrm{J}} = 82.5\ 10^{9} \qquad (8.10)$$

At 82 USD/barrel, this would have cost USD 6.765 10^{12} $\approx$ USD 6.8 trillion, which is more than the 6 trillion we would have to spend per year on building PV solar electricity generating capacity to replace fossil fuel consumption. Initially, we would have to spend 6 trillion in addition to the cost of fossil fuels. This is a big financial effort, but it is well within our means. As PV solar energy generation increases, expenditure on fossil fuels will decrease, being replaced by spending on electricity, which has largely already been paid for by the expense of installing the replacement PV solar energy capacity for fossil fuels.

The estimated expenditure is for unrefined crude oil at the wholesale price, which is significantly lower than what end consumers would have paid. Then, of course, not all the fossil energy came from crude oil; there was also coal and gas at different prices. Oil accounted for approximately 40% of total fossil fuel consumption. Regardless, treating all fossil fuels as crude oil still gives us an indicative number. In Chap. 9, we will come back to thinking in more detail about how to finance the energy transition.

8.4.1 Making Photovoltaic Solar Cells

Replacing fossil fuels with PV solar panels requires fabricating a large number of solar panels. Do we have the resources to fabricate that many panels? Can we produce the panels at the needed rate? What would be the environmental impact of their manufacture? To answer these questions, we need to examine the manufacturing of PV panels.

There are two main types of Photovoltaic (PV) solar cells: crystalline silicon and thin-film. Thin-film PV cells are cheaper to produce and require less silicon than crystalline PV cells, but have a somewhat lower efficiency (10–15%). By 2020, the market share of thin-film PV solar cells was around 5% of the total PV cell market. Research and Development of thin-film cells has intensified in recent years and is an ongoing work in progress. While this happens, we will focus on the fabrication of crystalline silicon PV cells, , which is the leading technology on the market.

To fabricate crystalline silicon cells, we need silicon of very high purity. Silicon (atomic number 14) is an abundant element on Earth and in the universe. It is the second most abundant element in the Earth's crust. However, silicon, like hydrogen, rarely occurs in its elemental form. In the Earth's crust, it occurs as silicon oxides, commonly referred to as silicates. Silicon dioxide (SiO_2) is commonly referred to as silica . Silica occurs abundantly in the Earth's crust as *quartz*,, which is the crystalline form of silica.

Silica is the raw material for crystalline silicon cells. To obtain silicon requires removing the oxygen from quartz by carbon arc welding of crushed quartz at temperatures between 1500 and 2000 °C, according to the chemical reaction equation

$$SiO_2 + 2\,C = Si + 2\,CO \qquad (8.11)$$

The result is pure (96% or above) *metallurgical* grade silicon and carbon monoxide. The carbon monoxide is further oxidised into CO_2 in an additional step (Maurits 2014). Metallurgical grade (MG) silicon is polycrystalline and not pure enough for solar cell production. Therefore, MG silicon, also known as silicon metal, must undergo further purification.

Solar-grade polysilicon requires a purity of 99.99999% (7N) or above. The *floating zone* technique (FZT) achieves this level of purity. It involves moving a silicon rod through a heated tube, where a small segment of the rod melts. As the melted zone advances, the impurities migrate with it towards the end of the rod, which is then cut off. The resulting ingot is still polycrystalline. It is recrystallised into a single crystal (monocrystal) using the Czochralski method, where the pure polysilicon ingot is melted again, and a monocrystalline ingot is pulled out of the melt. The final ingot is sliced into wafers that will become the solar cells. The wafers are *doped* with appropriate impurities to convert the silicon into a *p* or *n* type semiconductor. Before the wafer becomes an effective solar cell, it must undergo several chemical cleaning and etching steps and receive an anti-reflective coating. In summary, there are five stages in the production of PV panels:

1. Polysilicon production
2. silicon ingot fabrication
3. silicon wafer fabrication
4. solar cell fabrication
5. solar panel assembly

8.4.2 Fabrication Energy

The solar panel manufacturing process consumes large amounts of energy, primarily due to repeated silicon melting. Energy is one of the primary resources required for manufacturing PV solar panels, so it is helpful to know the amount needed. The solar cell must repay the energy used in its fabrication before it generates a net energy gain from the sun. A solar panel consists of solar cells, a supporting aluminium frame, and a glass cover. Solar cell manufacturing consumes about 70% of the total energy; the rest goes into making the aluminium frame and the glass cover.

Accounting for the energy required to manufacture a solar panel is quite challenging, and there are few published studies on this topic. The values published for the energy needed to manufacture a solar panel vary greatly. Estimated energy payback times range from a couple of years to more than 10 years, and some even claim that the panel can never recoup the energy consumed in its production. Let us investigate the truth.

The simplest way to estimate an upper bound on the energy required to fabricate a solar panel is to examine the panel's sales price and assume it equals the energy cost of its production. PV panels cannot be sold for less than their energy cost, or it would be a loss-making business. Of course, there are other costs, but let us ignore them for now to estimate an upper bound on the energy required to manufacture PV panels.

Let us start with the cost of USD 1.50 per Wp of installed solar panels, as calculated by the NREL (USA) and mentioned earlier.[3] If the electricity cost is 20 USD cents per kWh, then 1.50 USD/Wp allows us to pay for 7.5 kWh (27 MJ) of electric energy per watt peak.

Previously, we estimated that 1 Wp delivers 0.193 W end-user energy (see Eq. 8.6). Thus, the total average end-user energy per Wp over a twenty-year lifetime of a panel is

$$0.193\ \text{W} \cdot 3600 \cdot 24 \cdot 365 \cdot 20\ \text{s} = 121.7\ \text{MJ} \qquad (8.12)$$

[3] In 2022, the New Zealand retail price for installed domestic solar panels was 1.45 USD/Wp.

Subtracting the 27 MJ per WP spent in manufacturing the solar panel leaves us with a gain of (121.7–27) MJ = 94.7 MJ per WP. In other words, over 20 years, the panel pays back the energy spent in its manufacture 94.7/27 = 3.5 times.

The above fabrication energy estimate is an overestimate because the retail price also has to cover other costs, such as equipment amortisation, labour, and a profit margin at each production stage. The profit margin is probably the most important aspect. A modest net profit margin of 30% would result in an energy input of around 19 MJ per Wp and an energy payback of 5.4 times the manufacturing energy expenditure. On the other hand, we may have set the electricity price too high. Industrial electricity prices are usually lower than domestic prices. In this case, more energy could be paid for, giving us a higher energy cost for solar panel production. Adding the other production costs, which we ignored so far, leaves less money for buying energy, which has the same effect as overestimating the electricity price.

In 2015, Dai-Pra et al. (2015) analysed the energy payback times of solar panels, surveying the literature and finding an energy cost for solar panel fabrication of 23.54 MJ/Wp, which is only 13% lower than our upper-bound estimate. Despite uncertainty about the energy required for solar panel manufacturing, it is clear that, over their lifetimes, PV solar panels produce much more energy than they consume during manufacturing.

8.4.2.1 A Lower Bound Estimate

Another way to estimate the energy consumed in solar panel production is to calculate the energy used at each production stage. The most energy-demanding stages are the first three listed earlier, which comprise the fabrication of the solar cell. The process begins with the production of metallurgical silicon, which serves as the raw material for the fabrication of solar-grade polysilicon ingots. Recrystallising the polysilicon ingot into a solar-grade monocrystalline silicon ingot is the last of the high-energy-demanding steps. Adding the energy consumption of these steps yields an estimate of the minimum energy required for solar panel fabrication.

It takes 12 kWh (43.2 MJ) to produce one kilogram of silicon metal. It then requires an additional 70–120 kWh/kg (252–432 MJ/kg) to convert metal-grade silicon into solar-grade polysilicon (Chalamala 2018).

According to a 2002 survey by Jester (2002), the recrystallisation by an optimised Czochralski process requires approximately 40 kWh/kg (144 MJ/kg).

Let us assume it takes 43.2 + 350 + 144 = 537.2 MJ to produce one kilogram of PV wafers. Using our earlier estimate that we need 0.75 kg of solar-grade polysilicon per square meter of solar cell, the energy cost per m^2 is

$$537.2\ \frac{\mathrm{MJ}}{\mathrm{kg}} \cdot 0.75\ \frac{\mathrm{kg}}{\mathrm{m}^2} = 402.9\ \frac{\mathrm{MJ}}{\mathrm{m}^2} \tag{8.13}$$

or per Wp

$$\frac{402.9\ \frac{\mathrm{MJ}}{\mathrm{m}^2}}{180\ \frac{\mathrm{Wp}}{\mathrm{m}^2}} = 2.245\ \frac{\mathrm{MJ}}{\mathrm{Wp}}. \tag{8.14}$$

Solar cell fabrication accounts for approximately 70% of panel manufacturing costs. 2.24 MJ/Wp is 70% of 3.2 MJ/Wp, which is approximately one-eighth of our estimate based on the sales price. Although this lower bound is unexpectedly low, it is still of the same order of magnitude as our upper bound, suggesting that the manufacturing energy is of the order of at most a few tens of MJ per Wp.

8.4.3 CO_2 Emission Cost

The situation is less favourable for CO_2 emissions from solar panel manufacturing when fossil fuels are used to manufacture them. Using fossil-fuel electricity to produce the solar panels would require a fossil-fuel input to the (40% efficient) power plant of 27 MJ/0.4 = 67.5 MJ per Wp. Dividing by the *toe* makes 0.0016 tonnes (1.6 kg) of oil, which would emit 4.8 kg (three times the weight) of CO_2 per Wp. From this, we can calculate the emission per unit energy by dividing the 4.8 kg of CO_2 emitted during fabrication by the energy generated from 1 Wp worth of solar panel over its 20-year lifetime

$$\frac{4800 \frac{\text{g}}{\text{Wp}}}{0.193 \frac{\text{W}}{\text{Wp}} \cdot 3600 \cdot 24 \cdot 365 \cdot 20\,\text{s}} = 4.04\ 10^{-5} \frac{\text{g}}{\text{W s}}$$

$$= 40.4 \frac{\text{g}}{\text{MJ}} \quad (8.15)$$

This amount is about 80% of the 49.4 g/MJ of CO_2 emitted when burning methane. Therefore, solar cells are not a zero-carbon energy source when manufactured using fossil fuels. Only by using zero-carbon energy sources in solar cell manufacturing can PV solar energy become a zero-carbon energy source.

8.4.4 Production Capacity

Is there sufficient silicon available, and do we have the capacity to fabricate the required number of solar panels within the available timeframe? According to the IEA, worldwide solar panel production capacity increased by 386 GWp in 2023 to a total of between 700 and 800 GWp (Mason et al. 2024). Clearly, the existing solar panel production capacity is far too small for a fast-track electrification of the world's energy supply.

In Eq. 7.20), we estimated the total electricity demand for 2023 at 41.24 TWp. To spread the installation of this capacity over ten years makes an installation rate of 4 TWp/year, which is

$$\frac{4\,\mathrm{TW}}{386\,\mathrm{GW}} = \frac{4\;10^{12}\;\mathrm{W}}{386\;10^{9}\;\mathrm{W}} = 0.0104\;10^{3} = 10.4 \quad (8.16)$$

times the 2023 increase in world solar panel production capacity of 386 GWp, as quoted by the IEA.

To assess the feasibility of producing 4 TWp/year of solar panels, let us begin by estimating the raw materials required to produce 4 TWp. A crystalline PV solar cell is mainly a thin wafer, around 160 microns thick, sliced from a solar-grade (SOG) monocrystalline silicon ingot. These ingots are made from SOG polysilicon ingots by recrystallising the polysilicon using the Czochralski method. There is no substantial loss of material in recrystallising the polysilicon, but slicing the crystalline ingots into wafers wastes about half of the material. Then, how much SOG crystalline silicon do we need to make a 1 m^2 solar panel?

Because polysilicon production is measured by weight (mass), we first have to find the mass of the silicon wafers. With a wafer thickness of 160 microns, a square metre of wafers has a volume of 160 $10^{-6}m^3$. The density of crystalline silicon is 2228 kg/m^3; therefore, the mass (density times volume) of one square metre of wafers is

$$2228\,\frac{\mathrm{kg}}{\mathrm{m}^3} \cdot 160\;10^{-6}\;\mathrm{m}^3 = 0.35648\;\mathrm{kg} \quad (8.17)$$

Since half of the ingot is wasted during slicing, we need twice as much, namely about 0.75 kg of solar-grade crystalline silicon per square meter of solar panel.

In Eq. (8.8) we estimated that we need $2.78\ 10^5\ \text{km}^2 = 2.78\ 10^{11}\ \text{m}^2$ of solar panels. Because that result included $0.25\ \text{m}^2$ extra space around each panel, we have to divide this area by 1.25 (or multiply by 0.8). Thus we need

$$\begin{aligned} 2.78\ 10^{11}\ \text{m}^2\ \cdot 0.8 \cdot 0.75 \frac{\text{kg}}{\text{m}^2} &= 1.668\ 10^{11}\ \text{kg} \\ &= 167\ 10^6\ \text{tonnes} \end{aligned} \tag{8.18}$$

of solar-grade crystalline silicon ingots. Spread over ten years, this amounts to approximately 16.7 million tonnes per year. Therefore, assuming that there is no substantial material loss in refining metallurgical-grade silicon to solar-grade polysilicon and then to monocrystalline silicon ingots, we require 17 Mt (mega tonnes) of metallurgical-grade silicon per year.

According to Bernreuter (2022), in 2021, the ten largest producers had a combined capacity of 867800 tonnes (0.868 Mt). We need almost 20 times as much. This is roughly consistent with the factor of 10 that we estimated earlier for the required solar cell production capacity, and it reinforces the fact that solar-grade polysilicon production is utterly insufficient. Competition between manufacturers, tariff battles, and a lack of clarity on the market's future do not help increase production.

The required big increase in production capacity, even if intended, cannot be achieved in one year. It will take several years to ramp up production. A further complication is that once the total peak generation capacity has been installed, production will have to be cut back to the replacement level. With a useful life of 20 years for solar panels, the yearly replacement is 40 TWp/20 years = 2 TWp per year. That is half of the annual production needed during the 10-year installation period. Clearly, this requires some careful production planning!

Even if we could build up the required production capacity, we still need to ensure that the necessary resources are available. If we are to use monocrystalline solar cells, we require a sufficient supply of SOG polysilicon,, which in turn necessitates a supply of metallurgical-grade silicon. Where are the metallurgical-grade silicon producers, and what inputs do they need?

8.4.5 Metallurgical Grade Silicon Supply

Silicon is produced in two forms: metallurgical grade silicon and ferrosilicon. Ferrosilicon contains iron in various proportions and is used in the production of steel. Metallurgical-grade silicon, also known as silicon metal, is 99% silicon and is predominantly used in the metallurgical and chemical industries. In 2021, global silicon production was about 8.5 million tonnes (Statista), out of which around 2.5 million tonnes was silicon metal. About 15%, or 0.3 megatonnes, of this was used for solar cells and electronics (Chalamala 2018). However, we need 17 Mt per year of it! Producing the silicon required for PV solar panel fabrication will require a gigantic global infrastructure program. To illustrate the point, in 2021, the cost of a 10,000-tonne polysilicon plant with a capacity of 10,000 tonnes was approximately USD 500 million. We will need $17\ 10^6/1\ 10^5 = 170$ of those plants, costing USD 85 billion.

8.4.5.1 Supply Monopoly

China has become a near-monopoly supplier of polysilicon and, even more so, of wafers (approximately 98% in 2022). China is becoming a major hurdle for the fabrication of crystalline solar cells, as it has pricing power to keep competitors out of the market, thereby becoming a bottleneck. China has also become the dominant manufacturer of solar panel fabrication machinery. China appears to be prepared to limit the export of such machinery to consolidate its monopoly on solar panel fabrication.

8.4.6 Effect of Solar Panels on the Radiation Balance

We perceive solar panels as very dark, if not completely black, as they should be. They are designed to capture as much of the incoming solar radiation as possible by reflecting as little as possible. This raises the question of whether solar panels will alter the radiative energy balance of the Earth's surface they cover. To answer this question, let's examine the overall energy balance.

Albedo is the fraction of incident solar radiation reflected from a surface. In addition to the reflection of directly incident radiation, albedo also includes the reflection of diffuse light from all directions. The albedo depends on the surface type, orientation, solar angle (time of day), and wavelength. The surface of charcoal has an albedo close to zero. The albedo of the open ocean is as low as 0.06, and the albedo of forests can be anywhere between 0.08 and 0.18. The albedo of open soil is approximately 0.17, while that of new concrete is around 0.5, and that of fresh snow is approximately 0.8.

The average albedo of the Earth, as seen from space, is around 0.3. This value is surprisingly high, considering that 71% of the Earth's surface is ocean, which has a very low albedo. The high average albedo is due to clouds with their high reflectivity.

When examining the effect of solar panels, we will consider the case of clear sky. The albedo of a good solar panel is low, similar to that of the open ocean. Therefore, a panel absorbs nearly all of the incident radiation. With an efficiency of, say, 20%, the solar panel will convert 20% of the incoming radiation into electricity. The rest will be returned to the environment as heat. In that sense, the surface of a solar panel is like any other surface that has a low reflectivity. The surface of the oceans absorbs nearly all of the incident solar energy. In solar panels, the electricity generated by the solar cells often goes elsewhere, where the energy it carries will finally be turned into heat. Therefore, the *local* effect is the same as that of a surface with an albedo of around 0.2, which is that of a patch of open soil or vegetation. Only for solar panels installed in a high-albedo region, such as a desert, would it make a difference to the local climate. The climate effect would be similar to reforesting the desert.

The electric energy from the solar panels carried away for use elsewhere ultimately heats the environment, in addition to the local solar irradiation. But it will do less so than burning fossil fuels due to the higher efficiency of electrical equipment.

To return to our initial question, we can compare the amount of thermal energy added to the atmosphere from energy consumption to the amount of energy radiated from the ground and absorbed by greenhouse gases.

According to NASA Earth Observatory (Lindsey 2012), 17% of the total energy reaching the Earth radiates back from the surface. Part of this, 5% of incoming radiation, is captured by greenhouse gases in the atmosphere. The average rate of energy (power) reaching the top of the Earth's atmosphere is 340 W/m^2, which is 1/4 of the normal solar irradiation of 1361 W/m^2. That means greenhouse gasescapture

$6.8\,\mathrm{W/m^2}$. For the whole of the Earth's surface, this amounts to

$$510.06\ 10^6\,\mathrm{km}^2\ 6.8\frac{\mathrm{W}}{\mathrm{m}^2}\ 10^6\frac{\mathrm{m}^2}{\mathrm{km}^2} = 3468.408\ 10^{12}\,\mathrm{W} = 3468.4\,\mathrm{TW} \quad (8.19)$$

The average 2023 fossil fuel power consumption of 15.92 TW is a tiny 0.46% of that. It will even be somewhat less when fossil fuels are replaced by solar electricity, assuming that the concentration of CO_2 in the atmosphere remains constant thereafter. Thus, the answer to our initial question is that the contribution to atmospheric warming from the solar energy absorbed by the solar panels is negligible.

8.5 Wind

Ship sails are humankind's oldest invention for converting solar energy into mechanical work. The invention of the windmill then followed by the end of the first millennium CE. Nowadays, we attach electricity generators to windmills and call them wind turbines. Wind turbines are a mature technology for generating electricity without CO_2 emissions. It costs as much, or even less, to generate electricity with wind turbines than with fossil-fuel power stations. Therefore, wind energy appears to be a viable alternative to fossil fuel energy. To know whether we can scale up wind energy generation to cover world energy consumption, we must answer similar questions as we did for photovoltaic solar electricity:

- Can current wind turbines produce the needed energy?
- Do we have space for the turbines?

- Can we pay for the fabrication and installation of the turbines?
- Do we have the resources to fabricate the turbines?
- Are there any harmful side effects?

8.5.1 Energy Generation Capacity of Wind Turbines

Let us start by estimating the amount of energy available in wind . The source of wind energy is the kinetic energy carried by the moving air masses . Take a cubic metre of air. At normal atmospheric pressure and 20 °C, a cubic metre of air has a mass m of 1.28 kg. A fresh breeze has a speed of v = 10 m/s (36 km/h or 19.4 knots). Thus, the kinetic energy carried by one cubic metre of air in a fresh breeze is

$$E = \frac{1}{2}\mathrm{m}v^2 = \frac{1}{2}\dot{1}.28\frac{\mathrm{kg}}{\mathrm{m}^3} \cdot 100\frac{\mathrm{m}^2}{\mathrm{s}^2} = 64.28\frac{\mathrm{J}}{\mathrm{m}^3} \tag{8.20}$$

That is not much. It takes almost nine thousand times as much energy (568 kJ) as it takes to bring the water in a hot water kettle to the boil. On the other hand, there are lots of cubic metres of air circulating in the atmosphere around the Earth. The mass m_a of the atmosphere is large (5.27 10^{18} kg). If a fresh breeze were to blow everywhere, then the total kinetic energy E_k in the atmosphere would be

$$\begin{aligned} E_k &= \frac{1}{2}m_a v^2 = \frac{1}{2} \cdot 5.27\ 10^{18}\mathrm{kg} \cdot 100\frac{\mathrm{m}^2}{\mathrm{s}^2} \\ &= 2.63\ 10^{20}\mathrm{J} = 263\,\mathrm{EJ} \end{aligned} \tag{8.21}$$

This is approximately half of global fossil fuel energy consumption in 2023.

Clearly, this simple model of air circulating around the world at uniform speed is very different from how air moves in the atmosphere. Even though there are the *Jet Streams* of fast-moving air in the upper troposphere that meander around the globe from west to east, the air movement in the atmosphere, as we know from the wind, is far from regular. The movement is very turbulent and follows complex patterns. There are regions of high pressure where there is little wind, and regions of low pressure into which air flows at 30 m/s or more, making for stormy weather. Nevertheless, assuming a uniform fresh breeze around the world allows us to estimate the order of magnitude of the kinetic energy stored in the atmosphere at any one time. As our estimate shows, that energy is substantial.

One thing we observe is that the atmosphere's kinetic energy is in long-term equilibrium. The solar energy that drives the air circulation does not make the wind grow stronger forever, nor does it let it subside. If we extract some of the wind energy, we can expect the sun to restore it.

So, how much of that energy can we extract? The tool we use to extract energy from the wind is a windmill. It is a time-proven machine that converts wind into mechanical work. Windmills use rotating vanes, or sails, to capture some of the kinetic energy of the wind that flows through the area swept by the vanes. Contemporary wind turbines, that is, windmills connected to an electric generator, still work in the same way as windmills from yesteryear. The main difference lies in the shape of the vanes. A modern wind turbine typically has three slim *blades* that have the aerodynamic shape of an aeroplane wing. As the air flows around the blade, the difference in air pressure between the two sides of the blade produces a force, like the lift generated by an airplane wing, that pulls the blades to rotate.

The principle of operation of the wind turbine appears simple, yet the competing influences of many factors complicate the design of efficient wind turbines. For example, even if the wind incidence is perpendicular to the blades' plane of rotation, the airflow incident on the blades is oblique because the blades rotate. Furthermore, the blades' speed increases towards the tips. For balance, a turbine needs at least two blades. Adding more blades increases mechanical complexity, with diminishing marginal increases in efficiency. That is why practical wind turbine design has settled on three blades. Fortunately, we do not need to delve deeply into wind turbine design to understand the underlying energy balance. Basic physics and the principle of the conservation of energy suffice.

Irrespective of the complexity of the airflow around the turbine blades, what essentially happens is that the air slows down as it propels the rotating blades. A simple way to describe the situation is that a mass of air with an incoming speed v_i passes through the plane of the rotating blades and emerges on the other side with a lower outgoing speed v_o. The kinetic energy lost by the air mass is transferred to the blades, which rotate and crank an electric generator. What we want to know is how much of the wind's kinetic energy we can extract with a windmill.

Let us consider a time interval Δt. During this time, a mass Δm of air passes through the area A swept by the blades of the turbine. Before passing through the plane of the rotating turbine blades, the incoming air mass had an initial speed v_i and therewith the kinetic energy

$$E_i = \frac{1}{2}\Delta m \cdot v_i^2 \tag{8.22}$$

after passing through the turbine area A, the outgoing speed of the air mass has reduced to v_o, and the remaining kinetic energy is

$$E_o = \frac{1}{2}\Delta m \cdot v_o^2 \tag{8.23}$$

The energy extracted from the air mass is the difference between the kinetic energy E_i of the incoming air and the kinetic energy E_o of the outgoing air

$$\Delta E_e = E_i - E_o = \frac{1}{2}\Delta m \cdot (v_i^2 - v_o^2) \tag{8.24}$$

The power extracted from the wind is the energy extracted in the time interval Δt divided by the time interval

$$\frac{\Delta E_e}{\Delta t} = \frac{1}{2}\frac{\Delta m}{\Delta t}(v_i^2 - v_o^2) \tag{8.25}$$

Because the mass of air that passes through the turbine depends on the wind speed (and the area swept by the blades, which is constant), we want to express Δm in terms of the wind speed. The air mass travels at a speed v_i towards the turbine. During the time interval Δt, all the air up to a distance $d_i = v_i \cdot \Delta t$ from the turbine reaches the turbine. Thus, the volume ΔV of the air that passes through the area A swept by the turbine blades is

$$\Delta V = A \cdot d_i = A \cdot v_i \cdot \Delta t \tag{8.26}$$

and the mass is the volume times the density

$$\Delta m = \rho \cdot \Delta V = \rho \cdot A \cdot v_i \cdot \Delta t \tag{8.27}$$

$$\frac{\Delta m}{\Delta t} = \rho \cdot A \cdot v_i \tag{8.28}$$

Hence, we find that the power P_e extracted from the wind is

$$P_e = \frac{\Delta E_e}{\Delta t} \tag{8.29}$$

$$= \frac{1}{2}\frac{\Delta m \cdot (v_i^2 - v_o^2)}{\Delta t} \tag{8.30}$$

$$= \frac{1}{2}\rho \cdot A \cdot v_i \cdot (v_i^2 - v_o^2) \tag{8.31}$$

$$= \frac{1}{2}\rho \cdot A \cdot v_i^3 \; (1 - (\frac{v_o}{v_i})^2) \tag{8.32}$$

The power extracted from the wind is proportional to the cube (!) of the incoming wind speed multiplied by an *efficiency* factor

$$\beta = (1 - (\frac{v_o}{v_i})^2) \tag{8.33}$$

that depends on the ratio of the outgoing speed to the incoming speed, which, in turn, depends on the construction of the wind turbine. For $v_o = 0$, $\beta = 1$ and for $v_o = v_i$, $\beta = 0$. The case $v_o = 0$ is physically impossible because it implies that the air stops behind the blades, blocking further air from entering. Therefore, β can never be one.

The whole process happens at constant atmospheric pressure; therefore, the volume of the air mass Δm is the same before and after it passes through the turbine, and the density ρ remains the same. Because of its lower speed, the air mass downwind (behind) from the turbine travels a shorter distance during the time interval Δt than it travels upwind before passing through the turbine. Downwind of the turbine, the air mass does not fit into a cylinder with the same radius (that of the blades) as upwind from the turbine, and therefore, the air mass has to spread out over a larger area such that the volume downwind equals the volume upwind.

The energy required to push the air behind the turbine prevents the turbine from extracting all the kinetic energy from the incoming air stream. Albert Baetz already showed in 1919 that a wind turbine can extract no more than 16/27 (0.5926) of the power in the wind. This important result is Baetz's law, , which is remarkably easy to prove. For the interested reader, I have included the proof in Appendix A. According to Baetz, we cannot extract more than 59% ($\beta = 0.59$) of the incoming energy, regardless of how good the windmill's design is. State-of-the-art wind turbines are highly refined machines that can extract almost 50% of the available energy.

Now that we know how much energy we can extract from the wind, we can put some numbers into Eq. (8.32) to get some estimates. The incoming power is proportional to the area swept by the blades and proportional to the cube of the wind speed. At double the wind speed, there is eight times as much power in the wind.

The rotor diameter of a state-of-the-art wind turbine can be nearly 200 m. While the trend has been to build turbines with ever larger diameters, let us use a diameter of 100 m for our estimate. The area swept by the blades is

$$A = \pi r^2 = 3.14 \cdot 50^2\,\mathrm{m}^2 = 7854\,\mathrm{m}^2 \tag{8.34}$$

At a wind speed of 10 m/s, the power available to the turbine is

$$\begin{aligned} P_i &= \frac{1}{2}\rho \cdot A \cdot v_i^3 = \frac{1.28}{2}\frac{\mathrm{kg}}{\mathrm{m}^3} \cdot 7854\,\mathrm{m}^2 \cdot 10^3\frac{\mathrm{m}^3}{\mathrm{s}^3} \\ &= 5.03\ 10^6\ \frac{\mathrm{kg\ m}^2}{\mathrm{s}^3} = 5\ \mathrm{MW} \end{aligned} \tag{8.35}$$

If the turbine could extract half of that power, it would deliver 2.5 MW.

By dividing the equation for the power by the area swept by the blade, we get the incoming energy flux, that is, power per unit area,

$$F_i = \frac{P_i}{A} = \frac{1}{2}\rho \cdot v_i^3 \qquad (8.36)$$

which we can compare with the energy fluxes in the Earth's energy balance (see Sect. 3.4). At a wind speed of 10 m/s, the wind energy flux is:

$$F_i = \frac{1}{2}1.28\frac{\text{kg}}{\text{m}^3} \cdot 10^3\ \frac{\text{m}^3}{\text{s}^3} = 640\ \frac{\text{W}}{\text{m}^2} \qquad (8.37)$$

This is about eight times the flux of the sunlight absorbed by the atmosphere and nearly double the flux absorbed by the atmosphere from the energy radiated from the Earth's surface.

In 2023, the world's energy consumption was 619.63 EJ, or 1.7 EJ per day, which is 0.65% of our earlier estimate of the wind energy in the atmosphere. Extracting 0.65% of this reservoir's energy per day seems sustainable. Thus, there is sufficient energy in the wind to meet all our energy needs. How many turbines would we need to provide for all our energy needs?

Suppose we can extract 50% of the 0.64 kW per square meter per windmill face area available at a wind speed of 10 m/s. How many square meters of windmill face area would we need to supply the 19.65 TW of power that the world consumed on average in 2023?

$$\frac{19.65\ 10^{12}\text{W}}{0.5 \cdot 0.64\ 10^3 \frac{\text{W}}{\text{m}^2}} = 62.88\ 10^9\text{m}^2 = 63\ 10^3\text{km}^2 \qquad (8.38)$$

To get that much turbine area, we need

$$\frac{62.88\ 10^3\text{km}^2}{7854\ \text{m}^2} = 8.0\ 10^6 \qquad (8.39)$$

wind turbines with 100m of blade diameter. How much land area do we need to install eight million such turbines?

Placing wind turbines too close to each other reduces their efficiency due to the air turbulence they create. They should be spaced 3–5 turbine diameters (Gupta 2016) apart. They can be arranged in rows, but the distance between rows must be at least 5 to 9 turbine diameters. Using the optimal spacing means that each of our wind turbines occupies a land area of $500\ \text{m} \cdot 900\ \text{m} = 450,000\ \text{m}^2 = 0.45\ \text{km}^2$. To install the 8 million wind turbines, we need $8\ 10^6 \cdot 0.45 = 3.6\ 10^6\ \text{km}^2$. To supply all the energy demand, we need 3.6 million km^2 of land, which is about half (47%) of Australia's land area, assuming that the wind blows day and night at 10 m/s. Note that this is 13 times the land area required to generate the same amount of energy as PV solar panels.

For the end of 2021, BP reported an installed wind power capacity of 824.9 GW. The actual wind generation quoted by BP was 1861.9 TWh, which is 25% (0.258) of the nominal capacity. Thus, we need four times the number of turbines calculated above, which brings the required land area to approximately twice that of Australia. That is a large area, but we might be able to distribute it across many wind farms worldwide, including offshore wind farms. Offshore wind farms have the advantage that the wind is stronger and steadier over the deep ocean. Offshore wind turbines are costly. BloombergNEF estimates that in 2023, floating wind turbines cost USD 10 million per MW, which is 10 USD per W (Farley 2023), compared to USD 1.5 per W for PV solar. New designs aim at lowering the cost. Floating turbines facing downwind will align with the wind. The increased noise of this arrangement is irrelevant for offshore locations. Placing two counterrotating turbines on a floating platform balances the torque that tends to tilt the single-turbine platform. Also, arrays of multiple smaller turbines (rotors) appear to be more efficient and easier to maintain. However, the fabrication and installation of wind turbines involve the manufacture of large, complex mechanical parts, as well as significant construction work that is more difficult to automate than PV solar panel installations. Therefore, the cost of wind turbines has increased rather than decreased.

The conclusion is that, if pushed, we could extract enough energy from the wind to cover all our energy needs. Fortunately, we do not need to, because we also have PV solar electricity. Electricity from wind is only cost-competitive with PV solar when produced on large-scale wind farms. For smaller-scale electricity generation, PV solar is less complex than wind and hence more flexible. Wind complements PV solar energy because it continues to blow even after the sun goes down and when the sky is overcast; thus, it can help bridge gaps in PV solar generation. Still, the wind supply, like solar irradiance, is not continuous. It is just that the wind follows another rhythm than the sun. Long periods of low wind happen.

8.6 Storing Solar Energy

PV solar electricity production is highly variable throughout the day due to weather conditions and is nonexistent at night. Periods of several days with low solar irradiation are also common in many regions of the world. The same applies to electricity generation from wind turbines. Although the wind can blow day and night, periods of low wind are common. The result is a mismatch between solar electricity production and demand. The transition to an electric economy is only feasible if the energy is always available when needed. Therefore, solar electric energy must be stored to fill production gaps. The installation of solar energy generation must go hand in hand with the installation of solar energy storage.

There are three ways to store electric energy:

- in batteries.
- as chemical energy in combustible fuels such as ammonia or hydrogen.
- as mechanical potential energy by pumping water into hydroelectric dams.

Proven technical solutions for large-scale electric energy storage exist for all three methods.

8.6.1 Batteries

Batteries store electric energy by moving electric charge onto two electrodes. Connecting the electrodes to a *load* discharges the battery and recovers the electric energy. The movement of the charge through the load makes an electric current that flows until the battery is completely discharged. Remember that current is the rate of charge flow. Therefore, the current multiplied by the time is the total charge that flowed during the time t. Some batteries are rechargeable, and others are not. Only rechargeable batteries are useful for storing solar electricity. Rechargeable battery technology has advanced rapidly in recent decades, with increasing energy densities and decreasing costs. Although specific energy and energy density for batteries are still very low when compared with various fuels, as shown in Fig. 7.1, they store enough electricity to provide useful energy storage.

The standard unit of measurement for the amount of charge stored in a fully charged battery is ampere-hours (Ah).

That is, how long can the battery supply a constant current? For example, a fully charged 4 Ah battery will supply 4 A for 1 h, or 1 A for 4 h, until completely discharged. A battery maintains a voltage (electric potential difference) between its two electrodes (positive and negative terminals). A good battery maintains its specified voltage for most of the discharge, dropping significantly only towards the end. From the Ah rating, we calculate the energy stored in the battery by multiplying the Ah by the battery's nominal voltage, since current times voltage equals power. Multiplying the power by the time gives the energy delivered.

$$P = i \cdot V \tag{8.40}$$

$$P \cdot t = i \cdot V \cdot t \tag{8.41}$$

$$E = (i \cdot t) \cdot V \tag{8.42}$$

The electrodes of a battery are the cathode and the anode. The cathode is the electrode from which the conventional current flows out of the battery (when discharging), labelled as the + terminal of the battery. The anode is the electrode into which the *conventional* current flows.[4]

There are many electrochemical reactions suitable for battery construction. They differ in storage capacity, cost, energy density, recharging times, safety, and, of course, their chemical ingredients. Lead-acid batteries have been in use for over a century and continue to perform reliably. As of 2025, the lithium-ion (Li-ion) battery is the most widely used battery type for portable electronic devices, appliances, power tools, light electric vehicles, and domestic energy storage. Yet, there is ongoing intensive research for better batteries for various purposes. Electric vehicles primarily need batteries with high energy densities and short charging times.

[4]The conventional current flow is the direction of positive charge flow, which is the opposite direction in which the electrons flow!.

For energy storage from large solar or wind farms, low cost is more important than high power density. Li-ion batteries are still being improved. For example, using silicon anodes instead of graphite anodes in Li-ion batteries doubles the specific energy from less than 0.9 MJ/kg to 1.8 MJ/kg while speeding up recharging (80% recharge in 10 min).

Iron-salt batteries have a lower power density than Li-ion batteries, but they are simpler to manufacture from readily available materials and are therefore cheaper. They may replace lithium batteries for bridging energy storage in solar and wind farms. Iron-air batteries are also a promising possibility. The working principle is the reversible oxidation of iron. Oxygen is taken from the air during discharge and returned to the air on charge. Their manufacturers claim to be able to fabricate iron-air batteries with twice the specific energy of lithium-ion batteries, capable of storing power for 100 h, lasting over 10,000 cycles, and costing one-tenth as much as a lithium-ion battery. While new battery types are being developed, lithium-ion batteries currently dominate the market.

Lithium-ion batteries are bundles of individual cells connected to provide the desired voltage and capacity. Lithium is the lightest metal; its atomic number is 3, and it is a solid at standard temperature and pressure. Lithium does not occur in its elemental form in nature because it is highly reactive and flammable in air. Still, Lithium compounds are abundant in low concentrations in the Earth's crust and oceans. Richer deposits of commercial value are rare. One of the commercially important natural occurrences of Lithium is in Spodumene, which is a lithium aluminium inosilicate, $LiAl(SiO_3)_2$. The Spodumene-containing ore is crushed and ground to separate the Spodumene from other minerals and impurities. The aim is to produce a spodumene concentrate containing 6% lithium as a raw material for further refinement in battery production. The feedstock for lithium battery production is lithium hydroxide or lithium carbonate.

Lithium easily loses its valence electron, becoming a positive ion. This property, combined with its low mass, makes it so useful for batteries. Lithium ions carry a charge that can be reversibly intercalated into appropriate cathode materials. Lithium-ion battery cathodes are made from lithium oxide compounds of cobalt, nickel, manganese, and aluminium. There are many variations in the composition of the cathode materials that affect their characteristics, including cost, energy storage capacity, recharging time, operating voltage, and safety. Cobalt-containing cathodes offer high specific energy. However, cobalt supply is considered a high-risk part of the supply chain for increased battery production. Therefore, the search is on for cobalt-free, cheaper, and more abundant cathode materials. One such material is Lithium ferro-phosphate (LFP). The downside is that LEP cathodes have lower energy density.

Lithium-ion batteries have about three times the life of lead-acid batteries. A Li-ion battery lasts from 15 to 20 years. It is unlikely that Li-ion batteries will become a waste problem because a carefully designed battery enables high recycling rates. Li-ion batteries are thus an established and viable technology for energy storage.

Panasonic first introduced the 18650 Li-ion cell in 1994. Nowadays, many manufacturers offer slightly larger 21700 Li-ion cells . The number identifies the physical dimensions of the cell. It is a cylindrical cell 21 mm in diameter and 70 mm long. The typical 21700 cell has a nominal voltage of 3.6V and a capacity of 4.0 Ah ($3.6V \cdot 4A \cdot 3600\,\mathrm{s} = 51.84\,\mathrm{kJ}$). The volume of the cell is 0.024l, which yields an energy density of 51.84 kJ/0.024 l = 2.14 MJ/l. When bundling the cylindrical cells into larger battery packs, some space remains between the cylinders. With hexagonal dense packing, the volume occupied is 10% larger than the cell volume. The weight of the cell is 68.2g, which gives it a specific energy of 0.76 MJ/kg. A rule of thumb for the amount of Lithium in a cell is to multiply its Ah rating by 0.3 to obtain the amount of Lithium in grams. According to this, a 21700 cell contains 1.2 g of Lithium. A typical electric vehicle battery has around 8000 cells, which makes almost 10 kg (9.6 kg) of Lithium.

Battery electric energy storage is already on a growth path for use in light passenger transport (electric vehicles), domestic solar energy storage, and backup storage for the electricity grid alongside pumped hydroelectricity. The question is: How much battery storage do we need?

Certainly, we need sufficient batteries to power the fleet of electric vehicles that will replace the ICE passenger car fleet, as well as for backup storage for the PV solar panels that can provide most of the domestic electricity consumption.

Let us first think about how much backup energy storage for PV solar electricity is reasonable. First, we need to store enough energy for overnight usage and also to bridge periods of low sunshine. Cloudy periods are often geographically localised and shifting. Therefore, it may be possible to transfer energy from a neighbouring region with temporary higher sunshine to an area with lower sunshine. All this complicates estimating the needed battery storage to guarantee a steady PV electricity supply. Would 10 d of supply suffice?

Let us perform an order-of-magnitude estimate of the battery storage required to supply 10 d of energy. We calculated earlier that a $1\,\mathrm{m}^2$ panel rated at 180 Wp generates, on average, 3 MJ per day. To give us 10 d of backup, we need to store 30 MJ in a battery. Let us take 21700 cells. The energy density of these cells is 2.14 MJ/l. To store 30 MJ, we need $30\,\mathrm{MJ}/(2.14\,\mathrm{MJ/l}) = 14.02\,\mathrm{l}$ of battery. The 21700 cells are cylindrical, each having a volume of 0.024 l. We need $14.02\,\mathrm{l}/0.024\,\mathrm{l} = 584$ of those cells. The cylindrical cells allow a two-dimensional, hexagonal dense packing with a packing factor of 0.906. Thus, the package of 584 cells takes up $(584 \cdot 0.024)/0.906 = 15.45$ l. Allowing some space for the interconnections, then the battery pack for storing 10 d of electricity supply for one 180 Wp (1 m^2) panel will be approximately the size of a parcel of 25 cm × 25 cm × 30 cm, or two standard 10 l buckets. That means we can comfortably accommodate a battery pack for storing 10 d of solar electricity supply underneath each panel. In 2024, we could buy a 4 Ah 21700 cell for around USD 2.00. At this price, the cost for a 10-day supply of battery storage comes to $584 \cdot 2$ USD $= 1168$ USD. The cost per watt peak is 1168 USD/180 Wp $= 4.5$ USD/Wp, which is 4.33 times the cost of the panel (at USD 1.50 per Wp)

How much battery storage would we need to back up global electricity consumption? We already estimated that replacing all (2013) fossil fuel consumption with PV solar electricity would require installing around 42 TWp. For backing up 10 d of consumption, we need a battery backup storage of 0.17 MJ per Wp (30MJ/180Wp). To back up the world's electricity consumption, we need to store

Again, we plan for a 10-year transition period, and we want to store half of the required backup energy in batteries. We would then need to add 0.344 EJ worth of battery storage per year.

In 2023, the IEA estimated the 2025 global Li-ion battery manufacturing capacity at 0.0143 EJ (3.97 TWh). This means that we have to increase manufacturing capacity by nearly 24 times.

How much Lithium would that require per year?

The rule of thumb is to multiply the Ah rating by 0.3 to get the amount of Lithium in grams. Therefore, we need to calculate the Ah rating from the energy to be stored.

The Ah for the battery storage of 0.344 EJ is

$$\frac{0.344\ 10^{18}\,\mathrm{J}}{3.6\,\mathrm{V}} = 9.56\ 10^{16}\,\mathrm{Ah} \tag{8.43}$$

We multiply this number by 0.3 to get the amount of Lithium, in grams, needed per year:

$$9.56\ 10^{16} \cdot 0.3 = 2.87\ 10^{16}\,\mathrm{g} = 28.7\,\mathrm{Gt} \tag{8.44}$$

In 2023, the probable global Lithium reserves were 0.026 Gt (U.S. Geological Survey 2023). That means we only have one thousand of the Lithium needed to fabricate the batteries for a reduced backup time (5 d) for the first year of the required PV solar cell production! Recycling lithium from already discarded batteries would not make a big difference. The bottom line is that although we have the proven technical capability to back up intermittent solar electricity with batteries, we cannot utilize it at the scale we need due to an insufficient supply of the key ingredient. Unless we develop new battery technology, batteries can only play a limited role in solar energy storage. Definitely, it cannot provide the storage capacity needed for utility-scale solar energy generation. Pumped hydro and its variants will have to come to the rescue, as well as less efficient chemical storage methods, such as those using ammonia and hydrogen.

The reward for the effort spent on the above calculations is the insight that, more than the fabrication of PV solar panels or wind turbines, it is the problem of storing solar electricity to bridge the gaps in electricity generation that requires our primary attention. Adequate backup electric energy storage is not only the most costly part (in the current situation) of the transition to an all-electric economy, but also the part where the solutions are technically and economically less certain.

8.6.1.1 Emissions from Battery Fabrication

The amount of CO_2 emitted during the fabrication of rechargeable batteries appears substantial; however, the estimates vary greatly. As with solar panel fabrication, CO_2 emissions come from manufacturing the battery cells and from the mining and purification of the battery's anode and cathode materials. In addition to the CO_2 emissions, sourcing these materials causes other significant environmental damage. Emissions from manufacturing energy consumption are easier to estimate than those from the materials supply chain. The variety of materials involved (graphite, lithium, nickel, cobalt, manganese, lithium-iron-phosphates), along with their different locations and circumstances of occurrence, makes it difficult to obtain a representative average. Recent estimates that include the supply chain of battery materials put the carbon footprint of battery manufacture at somewhere between 70 and 120 kg of carbon dioxide per kilowatt hour (Peiseler 2024), or 19.4 to 33.3 kg/MJ. Earlier, in Sect. 8.4.3, we found that CO_2 emission from solar panel fabrication is approximately 40.4 kg/MJ. Thus, emissions from battery manufacturing are similar to those from solar panel manufacturing. The emissions arise from the use of fossil fuels to power materials extraction and battery fabrication. Therefore, it is essential to use zero-emission energy at all steps in the solar electricity supply chain.

8.6.2 Gravity Storage

Pumped hydroelectric storage uses the potential energy of water in the Earth's gravitational field to store energy.

The potential energy increase from lifting 1 l of water (1 kg) by 1 m is

$$1\,\text{kg} \cdot 1\,\text{m} \cdot 9.8\frac{\text{m}}{\text{s}^2} = 9.8\,\text{J} \tag{8.45}$$

To store the 30 MJ required for a 10-day backup of a 1m^2 PV solar panel (180 Wp), we need to lift 1 l of water to

$$\frac{30\ 10^6\text{J}}{9.8\,\text{J/m}} = 3.06\ 10^6\text{m} \tag{8.46}$$

or 1000 l to 3 km or 20 cubic metres of water to a more reasonable height of 150 m. Pumped hydroelectricity, that is, pumping water into a hydroelectric dam, corresponds to the latter combination.

Instead of using water, we can lift a denser object, such as a large rock or a concrete block, to a lower height with an electric hoist, then let it sink back down when needed, powering an electric generator to recover the energy spent lifting the block. The density of concrete is about 2.5 tonnes/m^3. To store the 10-day backup energy, we need to lift a mass of 300 tonnes to a height of 10 m. Three hundred tonnes of concrete is equivalent to 120 cubic metres, which is approximately the volume of a concrete block measuring 5 m by 5 m by 5 m. Clearly, this is a much less practical alternative to the 20 l battery pack backup storage. However, such gravity storage may still be commercially viable in some situations, for example, by utilising old mine shafts or where space is not a limitation.

8.6.3 Thermal Energy Storage

In addition to storing energy in their gravitational potential energy, rocks can store significant amounts of thermal energy. The specific heat of water (4184 J/kg K) is significantly higher than that of other substances. However, its low boiling point limits its usefulness for storing heat. Materials useful for storing large amounts of thermal energy need to be stable over the widest possible temperature range. It is possible to heat rocks to hundreds of degrees without affecting their material structure. It turns out that clays have a large heat capacity of around 850 J/kg K. Refractory bricks can withstand heating up to 1700 degrees, and heating a 1 kg clay brick to 1000 °C stores around 0.85 MJ. A lithium battery stores around 0.76 MJ/kg. Clay bricks can store useful amounts of thermal energy for industrial processes. Conversion of thermal energy to electricity is less favourable, though, due to losses in the conversion process. The conversion requires a steam turbine that typically operates at around 45% efficiency.

According to the IEA,industry consumes a third of global energy, with heat generation accounting for three-quarters of that total. A staggering 90% of that heat comes from burning fossil fuels.

A so-called thermal battery mainly consists of bricks that can be heated to temperatures over 1000 °C by passing a current through wires laid around the bricks, just like heating bread in a toaster. The bricks, helped by extremely effective insulation, can then retain heat for days with only minimal losses. When it is needed, the heat can be released in controlled doses at variable temperatures by blowing air through channels in the bricks. Thermal batteries are cheaper to manufacture than electric ones that require cobalt or Lithium.

8.7 Solar Power Plants in Outer Space

While the day and night cycle, as well as the changing cloud cover, make photovoltaic electricity generation on the ground intermittent, a solar panel in orbit around the Earth can receive unobstructed solar irradiation and generate electricity at full power all the time, except for the brief period when it crosses the Earth's shadow. There would be no supply interruptions. In addition, more energy could be collected per unit area of solar panels because there would be no atmospheric absorption or reflection of solar radiation. Isaac Asimov first described the construction of a solar farm in space, beaming the energy down to Earth by microwaves, in a science fiction story published in 1941. It sounds like a beautiful solution to the intermittency problem.

Solar cell arrays are routinely put into orbit to power space stations and satellites. Additionally, small amounts of energy have been transmitted by microwaves on Earth between two points in line of sight. Yet the idea of space-based solar energy capture remains in the realm of science fiction for two reasons: First, transporting materials into space is too expensive to make the solar energy collected this way price-competitive with other established energy sources. Second, transmitting energy via microwaves at the scale of the Gigawatts produced by a typical ground-based power station poses formidable challenges that still need much research and development. Nevertheless, multiple feasibility studies are underway. With the introduction of reusable launch vehicles (rockets), the cost of putting equipment into orbit may come down drastically. Even with intense development, this technology will take decades to mature and become cost-competitive, if at all.

9

Transition to an All Electric Economy

A rapid transition away from fossil fuels is essential for our future. The most promising path forward is harnessing solar energy. Solar technologies—such as hydroelectric, wind, and photovoltaic—are already proven and scalable. They can provide electricity at costs competitive with fossil fuels. With these tools, we can replace fossil-fuel-powered machinery with electric alternatives. Nearly all domestic energy is already electrified, and industries are following suit. Only heavy mobile machinery and long-haul aviation remain out of reach. The world can shift to electricity as its main energy source, but what would the cost be? Some argue this shift will be expensive. In this chapter, we will examine the numbers and explore ways to finance the transition from a fossil-fuel economy to an all-electric economy.

J. Sitte, *Global Warming: Facts, Myths and Remedies*,
https://doi.org/10.1007/978-3-032-20862-0_9

9.1 Paying for the Transition to Electricity

Even if we can build all the solar panels, wind turbines, batteries, and power grid upgrades necessary for a rapid transition to an all-electric economy, how can we pay for it?

While nations could embark on radical national programs to electrify their economies, reaching the necessary political consensus is wishful thinking. A more realistic approach would be to charge a fee for the consumption of fossil fuels.

Burning a tonne of fossil fuel produces about three tonnes of CO_2. Since fossil fuel use is easy to monitor, the amount consumed directly indicates the CO_2 emitted. Adding a sales surcharge on fossil fuels is a simple way to charge a fee on these emissions.

9.1.1 Carbon Emission Fee

Anyone who dumps waste products into the environment may harm others. It seems fair to hold the originator of the waste accountable for the damages caused. A simple way to at least partially compensate others for the damage is to ask the waste originator to pay a fee for the damage and for the safe disposal of the waste. A fee imposes a cost that the originator may wish to reduce. When the originator is a business, the disposal cost becomes a business expense that the company may want to reduce by dumping less CO_2 into the atmosphere.

A carbon fee is a compensation for the future damage caused by current emissions. Because CO_2 accumulates in the atmosphere, current damage also comes from past emissions. Fairness dictates that polluters should pay according to their accumulated emissions. To do that would require a retroactive surcharge that would pose an administrative burden of nightmarish proportions. So we'd better forget such an idea.

Despite the apparent simplicity and fairness of a carbon fee, its implementation is fraught with difficulties. First, how big should the fee be? It should be fair compensation for the damage. The problem is that the magnitude of the damage is largely unknown. We have no entities that collect and dispose of CO_2, who can tell us how much disposal costs. Second, who should collect the fee? And finally, what should be done with the money collected from the fee?

Economists have proposed trading *emission rights* since the late 1960s (Coase 1988). An *Emission Trading Scheme* (ETS) creates a *market* for carbon credits. A carbon credit certificate typically allows the emission of one tonne of CO_2. The 1997 Kyoto Protocol laid the groundwork for a framework of carbon emission trading. In 2015, Article 6 of the Paris Agreement consolidated the framework for international carbon emissions trading. Today, politicians and economists accept emission trading as their preferred method of financing the cost of replacing fossil fuels.

An Emission Trading Scheme as proposed and practised, is a watered-down form of a fee on CO_2 emissions. First, it does not apply to all fossil fuel consumers. It only targets enterprises in selected branches of the economy. Second, it grants specific enterprises the right to free dumping of a discretionary amount of CO_2. Once allocated, the recipient can trade the entitlement provided by the carbon certificate. Under the Kyoto Protocol, industrialized countries were required to cap their emissions at 95% of their 1990 levels by 2012. Nations set industry-specific allowances to meet that goal. An ETS requires a regulatory government body that sets the amount of free emissions and enforces enterprises' compliance with the allocated emissions.

Enterprises that emit less than their allowance can sell unused *carbon credits* to those wishing to exceed their quotas.

Supply and demand set the price of these certificates. If free emissions are set too high, demand decreases and supply increases, resulting in lower prices. Low carbon credit prices reduce incentives to cut CO_2 emissions.

An ETS does not guarantee lower emissions. Emissions are capped by the allowed amount, not by trading credits. Trading may reduce emissions by imposing costs on those who exceed their allowance and allowing them to buy credits. It also rewards those who emit less by letting them sell excess credits. If certificate costs are lower than the cost of cutting emissions, enterprises will simply buy certificates to keep or raise emissions. Regardless, the amount of emission is set by the number of allocated certificates.

EFTs have spawned a carbon *offset* industry with the sole purpose of selling carbon credits. The reason behind the carbon offset is that if someone can demonstrate a reduction of CO_2 content in the atmosphere as a result of some activity, then he/she may be entitled to a carbon credit certificate for the amount of CO_2 extracted from the atmosphere. Someone else can then buy the certificates to *offset* her/his emissions. An example is planting forests for the purpose of selling carbon credits on the grounds that the carbon captured by forests compensates for industrial emissions. As a result, airlines can ask their customers to voluntarily pay for *offsetting* the carbon emissions caused by flying. The net emission from this procedure is zero, or, as they say, it is carbon-neutral. The process of reducing the CO_2 content must not be a process that would have occurred naturally anyway. For example, Brazil cannot claim carbon credits for the CO_2 captured by the existing Amazonian forests.

EFTs have strong critics (Kill 2024) for their potential shortcomings and their susceptibility to manipulation, such as when countries create carbon credits for CO_2 captured by existing forests. Such abuses generate revenue by encouraging more emissions rather than reducing them. In the rest of the chapter, we will analyse the much simpler option of funding the transition to an all-electric economy by a direct universal surcharge on fossil fuel prices. Whether such a financing model will be adopted is not our primary concern. Instead, it is a *thought experiment* that helps us to identify the forces at play and how they interact.

What distinguishes a carbon surcharge, or carbon tax, from an ETS is that all businesses or persons dumping CO_2 into the atmosphere must pay the surcharge. A carbon fee is an inherently *fair* charge because fossil fuel consumption is roughly proportional to GDP. Therefore, the more affluent countries will have to pay more per capita than the poorer countries. When governments take it upon themselves to collect the fee, it may be considered a form of taxation. The risk of treating it as a tax is that governments may treat it as just another source of revenue and divert it to other purposes.

9.1.2 Policies for Carbon Pricing

Whether implementing a carbon tax or an ETS scheme, it is necessary to set a *carbon price*. For an ETS, the simplest approach would be to let the market set prices through an auction, based on supply and demand.

For a carbon surcharge or tax, a fair carbon price would ensure the revenue collected from the carbon tax covers the cost of purchasing, installing, and operating the solar panels needed to replace fossil fuel consumption. The surcharge would not be a compensation for potential damages; instead, it would minimise future damages, regardless of their magnitude. This policy, let us call it Policy 1, mandates that all carbon tax proceeds be invested in new solar energy generation capacity.

To calculate the price of carbon emissions under Policy 1, we need to know the cost of replacing fossil fuels with solar energy. To simplify the calculation, we will neglect wind and use only PV solar energy. We have already done most of the necessary calculations in Chap. 8. As we measure fossil fuel consumption by its energy content, let's remember that the useful energy we can obtain from burning it is always less than its energy content. We expressed this in Eq. 7.19 with an average efficiency factor e for the conversion of the fossil fuel power consumption rate P_{FF} into end-user power P_e

$$P_e = e \cdot P_{FF} \tag{9.1}$$

We estimated that, in the past, the average efficiency e has been around 0.5. In Sect. 8.2.1, we found that we have to multiply the peak power rating P_p of a solar panel by an efficiency factor ϵ to get the average effective end-user power P_e delivered by the panel:

$$P_e = \epsilon \cdot P_p \tag{9.2}$$

For panels with an efficiency of 18%, we estimated $\epsilon = 0.193\,\mathrm{W}/W_p$. When replacing fossil fuels with PV solar power, we need to know the peak power rating of the PV panels that replace a given amount of fossil fuel power consumption. We obtain the answer from the condition that the effective power delivered by solar panels rated at peak power P_p must equal the effective power delivered by the fossil fuel consumption rate P_{FF}:

$$\epsilon \cdot P_p = e \cdot P_{FF} \tag{9.3}$$

$$P_p = \frac{e}{\epsilon} P_{FF} \tag{9.4}$$

By using this identity, we implicitly assume that the effective solar electric power is equal to the mechanical or heat power delivered by the fossil-fuelled machinery. This is not entirely accurate, but the efficiency of converting electricity into mechanical work and heat is very high. Our estimates for e and ϵ are rough and already have a high level of uncertainty. By assuming the complete conversion of electric energy to mechanical work and heat in all uses, we only make a slight overestimate that will not fundamentally change the conclusions we derive from using Eq. 9.4. With that caveat, by using the values for ϵ and e we estimated earlier, we find that to replace one Watt of fossil fuel consumption, we need

$$P_p = \frac{0.5}{0.193} \, 1\,\mathrm{W} = 2.59\,\mathrm{W} \tag{9.5}$$

of solar peak power capacity. At a cost of USD 1.50 per W_p it costs

$$2.59\ W_p \cdot 1.5\ \frac{\mathrm{USD}}{\mathrm{W}_p} = 3.855\,\mathrm{USD}. \tag{9.6}$$

to replace 1 W worth of fossil fuel consumption with solar electricity. This estimate does not include the inevitable cost of solar electric energy storage. We will continue to ignore this cost in the calculations that follow, but we will remember that we can always include it by simply increasing the estimate for the cost per watt-peak. A solar panel lasts for, say, 20 years. Thus, once we install 2.59 W_p of solar electricity generation, we replace 20 years of one Watt worth of fossil fuel power consumption. Over the 20 years (630,720,000 s), this amounts to a total energy of

$$1\mathrm{W} \cdot 630{,}720{,}000\,\mathrm{s} = 630.72\,\mathrm{MJ}. \tag{9.7}$$

The investment of USD 3.855 in solar electricity generation buys us 630.72 MJ (175.2 kWh), delivered over 20 years at 1 W, at a price of USD cent 2.2 per kWh, which is about 1/10 of the average household electricity price in industrialised nations in 2024. Now, let us ask what it will cost to generate 1 W over 20 years using fossil fuels? To answer this question, we first need to know how much mass of fossil fuel we need to burn to generate 630.73 MJ of useful energy. To convert 630.72 MJ worth of fossil fuel into fuel mass, we use the *toe* of 41.868 GJ per tonne (see Sect. 5.1.2) and find that we need to burn.

$$\frac{630.72\,\mathrm{MJ}}{41.868\frac{\mathrm{GJ}}{tonne}} = 0.0151\ tonne = 15.1\,\mathrm{kg} \tag{9.8}$$

of fossil fuel to give us 630.72 MJ of energy. Because we pay for fossil fuel per litre, we want to find the volume of 15.1 kg of fuel. The density of crude oil is around 0.83 kg per litre; therefore, 1 kg of crude oil is

$$\frac{1\,\mathrm{kg}}{0.83\ \frac{\mathrm{l}}{\mathrm{kg}}} = 1.20\,\mathrm{l} \tag{9.9}$$

Over 20 years, we will burn $1.2 \times 15.1 = 18.12$ litres of crude oil. At a price of USD 80 per barrel (159 l) of crude oil, which is around USD 0.5 per litre, we will spend USD 9.06 in fuel costs to generate 1 W over 20 years. The cost per kWh will be USD cent 5.17, which is over twice the cost of PV solar energy. Spending now USD 3.855 on a watt of PV solar power offers a substantial discount on energy cost (at constant prices, of course). Replacing the world's 2023 fossil fuel power consumption of 15.92 TW (see equation 7.18) with PV solar power at USD 3.855 per Watt would cost USD 61.4 trillion. Spreading this expense over the 20-year lifetime of the solar panels requires a yearly investment of USD 3 trillion (in 2023 constant dollars), equivalent to 2.9% of 2023 World GDP (USD 105.4 trillion). Could we raise that money with a fossil fuel tax?

We demand that the tax or surcharge on the purchase of 630.72 MJ worth of crude oil must cover the USD 3.855 cost of installing the replacement solar electricity generation capacity. Above, we calculated that to get 630.72 MJ, we must burn 18.12 litres of crude oil. The surcharge on the sale of 18.12 l of crude oil has to be USD 3.855, or USD 0.213 per litre. At a price of USD 0.5 per litre, that amounts to 43% tax. It is a big tax, but we must remember that it also serves as an advance payment for future electricity consumption, which makes future electricity prices cheap. The cost picture becomes less rosy if the surcharge also has to pay for the installation of backup storage, which may cost several times as much per watt as the solar panels or wind turbines.

9.1.3 The Tax on CO_2

So far, we have calculated the tax on the energy content of the fossil fuel and also on the volume of crude oil. But the quantity of interest for global warming is the amount of CO_2 dumped into the atmosphere. Therefore, emission trading schemes aim to express the tax in terms of the amount of CO_2 emitted. The problem is that there is no simple way to directly measure CO_2 emissions. We always have to estimate the amount of emitted CO_2 by calculations starting from a quantity we can measure, such as the amount of fossil fuel traded. Therefore, we need to convert our tax from dollars per unit of energy to dollars per unit of emitted CO_2 mass. There are other CO_2 emission sources, for example, land clearing. If we want to tax these activities, then estimating their CO_2 emissions provides a bridge for translating the fuel tax into an equivalent tax on emissions from other sources. In Eq. 5.5, we found that on average, the burning of liquid fossil fuels releases 72 μg of CO_2 per Joule, or 72 kg/GJ. A tax of USD 3.855 per 630.72 MJ is a tax of 6.11 USD per GJ, or per 72 kg of emitted CO_2, which makes

$$\frac{6.11\,\text{USD}}{72\,\text{kg}} = 0.0849\frac{\text{USD}}{\text{kg}} = 85\,\frac{\text{USD}}{tonne} \tag{9.10}$$

We conclude that the tax on CO_2 emission should be USD 85 per tonne.

9.1.4 The Reduction of Fossil Fuel Consumption

Had we levied the tax in 2023, we would have raised the USD 3 trillion needed to replace one twentieth (5%) of fossil fuel consumption with PV solar power. The newly installed solar power would reduce fossil fuel consumption by 5% in the following year. Consequently, tax revenue would decrease by 5%, assuming that yearly world energy consumption remains unchanged and that we managed to install all the PV solar power capacity from the 2023 tax revenue. This reduction means that in 2024, tax revenue would not have been sufficient to purchase the PV solar power capacity required to replace all fossil fuel consumption within 20 years. Thus, the tax revenue from Policy 1 is insufficient to replace our fossil fuel consumption within 20 years. To maintain the replacement rate, we would need to raise taxes annually to offset declining tax revenue from reduced consumption. Alternatively, we could charge consumers for their electricity consumption from PV solar energy. Notably, if the fossil fuel tax were completely offset by the installation of new PV solar power, as it would have been in the first year, the solar electricity would be essentially free for consumers. The shortfall in tax revenue in the second year is 5% of USD 3 trillion, or USD 150 billion. To compensate for this shortfall, we can sell the solar electricity generated by the solar capacity installed in the previous year. To cover the gap, we need to sell the electricity at 0.0443 USD/kWh (see Mathbox 9.1.4). This accounts for approximately one-fifth, or 20%, of the average electricity price paid by consumers in 2023.

Mathbox 9.1.4

Calculation of the solar electricity price

At USD 1.5 per W_p, the USD 3 trillion of tax collected during 2023 pays for 2 trillion ($2\ 10^{12}$) W_p new PV solar capacity, which will, on average, generate

$$2\ 10^{12} W_p \cdot 0.193 \frac{\mathrm{W}}{W_p} = 0.386\ 10^{12}\,\mathrm{W} \quad (9.11)$$

of effective power. In a year ($3.1536\ 10^7$ s) this amounts to a total energy

$$0.386\ 10^{12}\,\mathrm{W} \cdot 3.1536\ 10^7 = 1.217\ 10^{19}\,\mathrm{J} \quad (9.12)$$

Because we pay in kWh, let us convert the $1.217\ 10^{19}$ J to kWh by dividing by $3600 \cdot 1000 = 3.6\ 10^6$ J/kWh

$$\frac{1.217\ 10^{19}\mathrm{J}}{3.6\ 10^6 \frac{\mathrm{J}}{\mathrm{kWh}}} = 3.38\ 10^{12}\,\mathrm{kWh} \quad (9.13)$$

To raise USD 150 billion from the sale of 3.38 trillion kWh of solar electricity, we have to charge.

$$\frac{150\ 10^9\ \mathrm{USD}}{3.38\ 10^{12}\,\mathrm{kWh}} = 0.0443 \frac{\mathrm{USD}}{\mathrm{kWh}} \quad (9.14)$$

Thus, with a carbon tax of 0.231 USD per litre of crude oil and a price of 4.4 cents per kWh of PV solar electricity, we can finance the complete replacement of the world's 2023 fossil fuel consumption with PV solar electricity in 20 years. For comparison, the IEA has calculated that to have any hope of achieving net-zero by 2050 would require adding 630 gigawatts of solar photovoltaics and 390 GW of wind each year, starting no later than 2030. Which is about four times any annual tally so far (Kroposki and Hoke 2024). Furthermore, the replacement rate calculated by the IEA, at 630 GW + 390 GW = 1.02 TW, is only half of the replacement rate (41.24 TWp /20 $\approx$ 2 TWp) we calculated earlier (see Eq. 7.20)

9.1.5 Optimistic Guess of Future Temperature Rise

Current solar panel technology assures a useful lifetime of at least 20 years for crystalline silicon solar cells. Crystalline silicon solar cells do not suddenly die after 20 years; instead, their efficiency declines over time. After 20 years, they may retain around 80% of their peak power rating and can operate for many more years at reduced power. The implication is that it may take more than 20 years to generate the 630 MJ of energy calculated under the assumption of full-efficiency operation over the 20-year useful life. Thus, with the supplemented carbon tax scheme we described above, it will take more than 20 years to fully replace all fossil fuel consumption. Over the 20+ years until CO_2 emissions from fossil fuel consumption cease, we will emit a lot of CO_2 into the atmosphere, keeping global warming going. Let us calculate how much additional CO_2 we will be dumping into the atmosphere until fossil fuel consumption stops. This is easy to do if we stick to the simplified case where the solar panels operate at full efficiency for 20 years and assume that our energy consumption stays at the 2023 level. With the supplemented tax scheme, the fossil fuel consumption decreases linearly over time, reaching zero after 20 years. Therefore, the total amount of fossil fuel consumed in the 20-year transition period is half of 20 times the initial consumption. Initial consumption was 504.83 EJ. Ten times is 5048.3 EJ. At an emission of 72 kg of CO_2 per GJ the total CO_2 emission over the 20 years will be

$$5048.3\ 10^{18}\,\text{J} \cdot \frac{72\,\text{kg}}{10^9\,\text{J}} = 5048.3 \cdot 72\ 10^9 = 363.5\,\text{Gt} \tag{9.15}$$

At an estimated warming rate of 0.00141 ° C per Gt, 363.5 Gt means an additional temperature rise of 0.5 CO_2 C.

We can reduce emissions by raising the tax to, say, double the current amount, so that the yearly rate of fossil fuel replacement doubles and full replacement of fossil fuels takes only 10 years. To compensate for the yearly shortfall in tax revenue, we will have to double the price of the newly generated solar electricity. The benefit of doubling the tax is that the additional CO_2 we will be dumping into the atmosphere until fossil fuel consumption stops will be halved, and so will be the increase in temperature. The total cost, of course, remains the same.

9.1.6 Impact on Panel Production Capacity

There is a catch with doubling the tax. With more money to install new PV solar generation capacity, we also need the capacity to fabricate and install twice as many solar panels per year. After 10 years, when all required panels are installed, demand will drop because even the panels installed first will still have 10 years of useful life left. It is after 20 years that panels will start to need replacing. The number of panels to be replaced each year after 20 years is half the number of new panels installed each year during the initial 10-year replacement stage. The solar panel manufacturing industry will then face very low demand, which will stabilise in the long term (after 20 years). Thus, the solar panel manufacturing industry will be reluctant to invest in the building of production capacity that will only be sustainable for the next 10 years.

In Sect. 7.9 we calculated that to replace the 2024 fossil fuel consumption with PV solar electricity in 20 years, we need to install 2 TW_p per year. According to the Energy Institute's 2023 World Energy Statistics, solar generation capacity increased by 345.8 GW from 2022 to 2023, and wind generation capacity increased by 116 GW, for a combined increase of 461.8 GW. To reach the sustainable electricity generation installation capacity of 2 TW_p per year, we need to increase the current solar and wind electricity generation installation capacity by over 4 times (4.34). However, the IEA estimated that by 2024, the capacity for new PV solar electricity installations would reach around 1.1 TW (IEA 2023). If that is true, then, given the much lower demand indicated by the actual installed new capacity, the solar panel industry is already suffering from overcapacity. Alternatively, we can view this situation as an opportunity for a substantial reduction in fossil fuel consumption.

9.1.7 Implications of the Policy

The policy of investing all revenue from carbon pricing in PV solar power installations pays for the capital costs of PV solar power upfront. Consumers only have to pay for the low running cost. PV solar power will be very cheap. This frees up money for consumers to invest in replacing fossil-fuel-consuming equipment with electric equipment, taking advantage of low-cost solar power.

9.1.8 How to Collect the Carbon Pricing Revenue

The biggest polluters need to make the largest investment in PV solar energy. It stands to reason that PV generation must be installed where consumers of fossil fuels are located. Therefore, the carbon *cost* should be borne by consumers locally, and the revenue should also be invested in installing local PV solar electricity generation. This could be accomplished by a national authority commissioned to collect carbon-pricing revenue and oversee investment in PV solar generation capacity. Logistically, revenue collection could be done in a similar way to how goods and services taxes are collected in many countries.

9.2 Electric Cars

The transition to an all-electric economy seems simple, in principle. Yet it hides many potential traps that require careful analysis. The electrification of personal transport provides an insightful example. In the 1980s, air pollution from passenger cars garnered much public attention, prompting the introduction of emissions restrictions and fuel-efficiency targets. Later, as attention shifted to global warming from CO_2 emissions, passenger cars remained seen as a major cause. Thus, replacing the CO_2-spewing vehicles powered by internal combustion engines with electric *zero-emission* cars is seen as an urgent step towards overall CO_2 emission reduction. Let us see if this impression is supported by the facts.

9.2.1 Vehicle Emissions

By how much can we reduce CO_2 emissions by replacing an ICE vehicle with a battery-powered electric vehicle? The total CO_2 emission of any technical system is composed of two parts: manufacturing emissions and operating emissions. First, we will estimate operating emissions, as these are easier to estimate than manufacturing or production emissions. To estimate the amount of CO_2 emitted from running the average ICE vehicle, we will estimate the weight of the fossil fuel consumed per distance travelled. Once we know this, application of the rule of thumb (see Eq. 5.3) that the weight of CO_2 emitted is three times the weight of a liquid fossil fuel burned tells us the weight of CO_2 emitted per distance travelled. To estimate the weight of petrol burned by the average vehicle in a year, we need data on the average distance traveled per year and the average fuel consumption per unit distance. The Australian Bureau of Statistics reported that in 2019, the average fuel consumption of passenger vehicles was 11.1 l per 100 km, and the average distance travelled per year was 11,100 km.

The fuel consumption for ICE vehicles is given in litres per 100 kilometres. To translate fuel consumption volume directly to mass of CO_2 emitted per 100 km, we multiply the volume by the density of gasoline and then the resulting mass by three

$$\frac{\text{kg CO}_2\ emission}{100\,\text{km}} = \frac{litres\ fuel}{100\,\text{km}} \times \frac{\text{kg}}{\text{l}}(density) \times 3 \tag{9.16}$$

The density of gasoline is 0.75 kg/l, thus

$$\frac{\text{kg CO}_2\ emission}{100\,\text{km}} = \frac{1\ fuel}{100\,\text{km}} \times 2.25\frac{\text{kg}}{l} \tag{9.17}$$

For the average consumption of 11.1 l/100 km, this makes

$$\frac{11.1\text{l}}{100\ \text{km}} \times 2.25\frac{\text{kg}}{\text{l}} = 24.975\ \frac{\text{kg CO}_2}{100\ \text{km}} \tag{9.18}$$

On average, an ICE passenger vehicle emits 25 kg of CO_2 per 100 km. Over the year, in Australia, a passenger vehicle travelled an average of 11,100 km and emitted 2.77 tonnes of CO_2 in 2019. Multiplied by the 14.5 million passenger vehicles registered in Australia in 2019, the total emissions from passenger vehicles in Australia were 40.2 Mt of CO_2. This is 10% of the total 401 Mt of CO_2 emitted by Australia in 2019 (BritishPetroleum 2020). If we neglect the small number of electric passenger cars on the road in 2019, then our result is that the full replacement of Australia's ICE-powered passenger vehicle fleet by EVs could reduce the country's CO_2 emissions by 10%; with the proviso that the EVs are powered exclusively by renewable electricity. Although this is a substantial amount that should be achieved rather sooner than later, it is not "the" big contribution to the goal of reaching zero carbon emissions, as it is often assumed.

Having established the figures for Australia, we next consider whether these estimates can be applied elsewhere. Australia has one of the largest per capita kilometres travelled in passenger vehicles. Therefore, the estimates of Australia's savings in CO_2 emissions from replacing ICE passenger vehicles with electric passenger vehicles represent a high-end estimate for the world average.

Given the current state of affairs, where 84% of all energy is still obtained from fossil fuels, it is overly optimistic to assume that all EVs will be powered by electricity from renewable sources. Therefore, we should ask what the savings are from switching to EVs when they are powered by electricity generated from fossil fuels.

The key quantity needed to find the answer to this question is an estimate of the efficiency of converting the energy consumed into *transportation* work. Where transportation work means the distance traveled on the road. For ICE vehicles, the amount of fuel consumed per distance travelled under *normal* driving conditions is not difficult to measure and is readily available, as we have seen from the Australian Bureau of Statistics data. For EVs, the corresponding measurement would be the amount of electric energy consumed per distance travelled. The way to do this is to measure the electrical energy consumed at the power point during charging, and then measure the distance travelled by that amount of charging. There is a scarcity of data from independent sources, and we must rely on the claims made by EV manufacturers. Manufacturers quote kWh/100 km values ranging from around 13 at the lower end to up to 28 for upper-end luxury EVs, with many vehicles around 16 kWh/100 km. This is about 1/7-th of the 104 kWh/100 km average energy consumption for an ICE vehicle rated at 11.1 L/100 km. The high efficiency of EVs stems from the fact that, in addition to electric motors being more efficient than internal combustion engines, regenerative braking further enhances EV efficiency. In regenerative braking, the electric motor is switched to generator mode. The braking force is obtained from the torque required to turn the generator and produce electricity. The electricity generated from braking is then used to charge the battery.

To find the weight of fossil fuel burned by the electric power plant to produce the 16 kWh of electric energy consumed by the EV traveling 100 km, we note that state-of-the-art fossil fuel electric power plants operate at around

40% efficiency. Therefore, disregarding losses in the power distribution network, generating 16 kWh at the power outlet would require 16/0.4 = 40 kWh of energy input to the plant. Converting back to Joules, this makes 40 kWh $\cdot$ 3.6 MJ/kWh = 144 MJ. Dividing by the tonne of oil equivalent of 41.868 MJ/kg, we obtain 3.44 kg. Thus, an EV rated at 16 kWh/100 km will consume 3.44 kg of fossil at the power plant, resulting in an emission of 10.3 kg of CO_2 per 100 km. This is 41.3% of the 24.975 kg emitted by the average ICE vehicle. Thus, even when using fossil fuel electricity, an EV emits only 41% of the CO_2 emitted by an ICE vehicle. However, for the less efficient high-end luxury EVs with a consumption of around 28 kWh/100 km, that grows to 72%, making the emission reduction far less impressive.

One may think that in countries such as Norway and New Zealand, where a large fraction of electricity generation comes from renewable sources, transitioning to electric cars could completely eliminate CO_2 emissions from passenger cars. For this to be the case, the country needs sufficient reserve renewable electricity generation capacity to meet the additional energy demand resulting from shifting vehicle energy consumption from fossil fuels to renewable electricity. In New Zealand, for example, 82% of the electricity comes from renewable sources, mainly hydroelectricity. By switching from internal combustion engine vehicles to electric vehicles, the electricity grid now has to supply all the energy for the EVs. The fact that 82% of electricity generation comes from renewable sources tells us that there is no more capacity available, and the remaining 18% must be covered by fossil fuels. Therefore, even in New Zealand, the full potential reduction in CO_2 emissions from switching to EVs will not be realised if fossil fuels have to provide the marginal energy for electric consumption.

How large would the additional demand on NZ electricity generation be if the ICE vehicle fleet were replaced by EVs?

The size of the New Zealand light passenger car fleet in 2019 was 3.4 million vehicles. Lacking specific statistics for New Zealand, let us use the Australian value of 11,100 km as the average yearly distance traveled. Using the optimistic value of 16 kWh/100 km for the electric vehicle fleet, we arrive at the following estimate of

$$3.4\ 10^6 \cdot 11100\ \text{km} \cdot 16\ \text{kWh}/100\ \text{km} = 6038\ 10^6\ \text{kWh} = 6.04\ \text{TWh} \tag{9.19}$$

extra demand on the NZ electricity generation[1]. The 2019 NZ electricity generation was 44.6 TWh, out of which 82% (36.57 TWh) was from renewable sources. If the extra EV consumption were to be covered by renewable electricity, the renewable electricity capacity would have to increase by 16.5%. Of course, the ideal situation would be for this energy to be generated by the vehicle owner's rooftop PV solar panels.

What fraction of New Zealand's total CO_2 emission would be saved by switching to EVs? In 2019, New Zealand's CO_2 emission from burning fossil fuels was 38.4 Mt. At our estimated emission of 2.77 tonnes per vehicle per year, the total emission from the New Zealand ICE light passenger vehicle fleet adds up to $3.4\ 10^6 \cdot 2.77 = 9.418 Mt$ of CO_2 per year. This is 27.4% of the total emissions from burning fossil fuels that could be saved by replacing the passenger car fleet with EVs powered by renewable energy. The emissions saved would drop to 15.5% if the EVs were powered by grid electricity from fossil fuels. New Zealand's relative contribution to CO_2 emissions from light passenger vehicles is large because other energy uses have lower emissions, given the large fraction of available electricity already coming from renewable sources.

[1]On average, each vehicle would consume 1776 kWh per year. For comparison, the average electricity consumption per NZ household is 7000 kWh

9.2.2 Money Savings for the EV Owner

Let us estimate the fuel savings from running an EV in Australia. By driving the average 11,100 km per year, an EV rated at $16\,kWh/100\,km$ will consume 1776 kWh per year. At a cost of AUD 0.25 per kWh when charging from the grid, this amounts to AUD 444 per year. The average amount of fuel consumed by an ICE car in a year would have been 1232 litres of unleaded petrol. With petrol prices in 2019 around AUD 1.5 per litre, that would have cost AUD 1846 for the year. Thus, the average owner of an electric car would have saved AUD $1846 - 444 = 1402$ in fuel costs. Over a 10-year period, that would amount to around AUD 14,000. New EV purchase prices were substantially higher than those of their corresponding ICV vehicles. To achieve a payback period of 10 years, the EV's price must not exceed the cost of an ICE vehicle by more than AUD 14,000. Even in 2022, EVs sold at a price premium well above that. One must conclude that, based on the fuel cost savings over a 10-year operating period, the purchase of an electric vehicle in 2019 was not a profitable investment, assuming that fossil fuel prices would not increase over time.

The savings in fuel from switching to an EV would be greater with a carbon pricing policy in place, as described earlier. We estimated that the carbon tax would increase fuel costs by 42%, from AUD 1846 to AUD 2621.32 per year. The fuel saving per year would be AUD 2080. Over ten years, this amounts to AUD 20,800, which would come close to compensating for the higher purchase price of an EV compared to an ICE vehicle. The carbon price tilts the balance somewhat in favour of purchasing an EV.

For prospective EV owners, the economic case remains unclear. Therefore, in 2024, sales of battery-electric vehicles (BEVs) began to decline in some countries. Not only are EVs more expensive to buy than ICE vehicles, but home chargers also add to the costs, as do higher insurance premiums, and in some countries, cheaper petrol. Rapidly evolving technology also makes depreciation uncertain. High import duties imposed in the USA and Europe on lower-priced BEVs from China make BEVs more expensive in these regions. For some manufacturers, pure electric vehicles are even unprofitable. This seems to be one of the reasons behind the intense marketing of the more profitable hybrid vehicles.

9.2.3 EV Purchase Subsidies

To make the transition to EVs more economically attractive, governments in several countries have offered or are offering subsidies for the purchase of electric vehicles. These subsidies are either paid from general government tax income or from a carbon pricing scheme. If this subsidy is paid from general tax revenue, then it is an unfair levy on every taxpayer for the benefit of a few. One could argue, though, that it is a tax that benefits everyone, as reducing CO_2 emissions slows global warming for everyone.

The CO_2 emissions from light passenger ICE vehicles in most countries account for between 10 and 20% of total fossil fuel CO_2 emissions. To completely eliminate CO_2 emissions from the light passenger car fleet by replacing them with EVs, the electricity used to power these EVs must come entirely from renewable sources. Therefore, the first step towards replacing the ICE passenger car fleet with EVs is to ensure there is sufficient renewable electricity generation capacity to meet the EV fleet's energy needs. Therefore, any subsidy should be directed at increasing the generation of renewable electricity. One way to achieve this is to subsidize the installation of domestic PV solar panels and battery storage, which could be used for overnight EV battery charging, as well as the provision of charging stations in densely populated areas where residents do not have access to a private charging point.

Subsidising the acquisition of EVs directly, instead of solar electricity generation, is thus less effective and has some unwanted side effects. One side effect is that if renewable energy capacity is unavailable, the additional demand for electricity is likely to increase electricity prices. The other is that car manufacturers, in response to increased demand for EVs, will not miss the opportunity to raise vehicle prices. These two side effects will cancel part of the net benefit of the subsidy to the buyer and also penalise every other electricity consumer.

Subsidies at the supply side, where EV manufacturers obtain government subsidies on environmental grounds, are thus misguided. The resources should be invested in solar electricity generation capacity instead. In countries where car manufacturing is a significant contributor to the economy, the purpose of such subsidies may be to support the international competitiveness of their automobile industry.

9.2.4 Hybrid Vehicles

The push by manufacturers to overcome the range limitations of battery-powered electric cars by offering *Plug-in Hybrid* electric vehicles (PHEVS), particularly in the SUV class, annihilate any environmental and economic benefits. Hybrid vehicles have both an ICE motor and an electric motor as well as a battery and a fuel tank, which makes them heavier than either equivalent EVs or petrol cars, and therefore require more energy to run. To limit their weight, the batteries in hybrids are smaller than in a pure EV and thus have a comparatively lower electric drive range. If owners are not meticulous in keeping the car charged, they will drive more in petrol mode than in EV mode.

9.2.5 The CO_2 Emission Cost of Car Manufacture

Assessing the environmental benefit of EVs requires more than just counting operational CO_2 emission savings. Calculating the net environmental cost also requires considering CO_2 emissions from the manufacture and disposal of the EV. As we shall see, including manufacturing energy consumption drastically changes the naive view obtained by considering only operational (running) energy consumption. Calculating total CO_2 emissions from car manufacturing is complex due to a vast, interconnected supply chain using diverse resources. For the following analysis, I will use the results of the work of Qiao et al. (2017), who compared the CO_2 emissions from manufacturing ICE and electric vehicles in China. At that time, China produced a quarter of the world's passenger vehicles. According to their findings, manufacturing an ICE vehicle produced 9.173 tonnes of CO_2, while manufacturing an electric vehicle produced 14.643 tonnes. EV production emits 60% (59.6) more CO_2 than that of an ICE vehicle. Notably, battery production alone produced 2.892 tonnes (20%) of CO_2. These calculations used the energy sources prevalent in China at the time. The calculations of CO_2 emissions hinge on the energy source mix used in the production process. If, instead, renewable energy were used in all stages of manufacture, there would be no CO_2 emissions and the whole issue would go away.

To illustrate this further, Daimler AG conducted a full life-cycle assessment of its Mercedes-Benz EQC 400 EV. The study revealed that, of the total energy consumption of 478 GJ over the vehicle's entire life cycle, 64.

Assuming that an equivalent ICE vehicle would require 60% of that energy, then the ICE vehicle would run off the assembly line with 13.15 tonnes of CO_2 emissions to its account. At the average yearly emission of 2.77 tonnes, calculated earlier, we could run the ICE vehicle for 3.17 years before reaching the emissions already accumulated by the brand-new MB EQC 400. Looking at it from a different perspective, if I have a ten-year-old petrol passenger vehicle that produces 2.77 tonnes of CO_2 per year and I replace it with an MB ECQ 400, I would cause a release of 22 (21.92) tonnes of CO_2. I could run my old ICE vehicle for another 8 (7.93) years (practically to the end of its useful life) before reaching the same amount of emissions. This simple calculation again illustrates how misguided government subsidies for the purchase of electric vehicles are. Taking manufacturing CO_2 emissions into account drastically changes the picture.

Assessing the CO_2 emission difference between the manufacturing of an EV and an ICE vehicle is a complex exercise. Accordingly, there is a scarcity of data to aid the comparison. The primary distinction between the two types of vehicles lies in the drivetrain: in one case, an internal combustion engine, transmission, and fuel tank; in the other, an electric motor(s), a simpler transmission, and a battery. The magnets in the electric motors and the batteries require materials that are less abundant than the steel and aluminium used in the petrol engine. An EV has fewer moving parts than an ICE vehicle, making it more reliable and simpler to manufacture, except for the battery. The EV's useful life could be longer than that of an ICE vehicle. A significant difference may arise from the EV's longer lifespan, leading to fewer cars being manufactured over time.

9.3 Air Travel

The ever-growing air travel industry is one of the few sectors for which we do not yet have a proven technology to replace its dependence on fossil fuels. Aircraft for long-haul air transport, passenger and cargo, need to have enough energy storage on board to travel non-stop over several thousand kilometres. A simple calculation reveals that on-board batteries are insufficient to store the energy required for long-distance air transport.

A typical passenger flight from Hong Kong to Frankfurt in an Airbus A350 consumes around 92000 kg of fuel. With an energy density for Jet fuel A of 42.8 MJ/kg, the total energy consumption for the flight would be $42.8\ 10^6$ J/kg · $92\ 10^3$ kg = $3.94\ 10^{12}$ J. Data for the actual fuel consumption of commercial flights is hard to come by. Therefore, Burzlaff (2017) worked out a method of calculating it. The calculation is rather complicated. We can, however, make a simple estimate of fuel consumption based on the published average fuel consumption for an A350 of 7.07 kg/km (Wikipedia) for the 9152 km distance from Hong Kong to Frankfurt. The result is 64,704 kg of jet fuel, which is 30% less than Burzlaff's result but still of the correct order of magnitude.

Having calculated the energy required for the flight, we can now calculate the weight of a Li-ion battery capable of storing that amount of energy. The best energy density of Li-ion batteries is 250 Wh/kg (0.9 MJ/kg), which is 2.1% of the energy density of jet fuel. To store 3.94 TJ, we need $3.94\ 10^{12}$ J/0.9 10^6J/kg = $4.38\ 10^6$kg = 4380 tonnes of battery, which is absurd given that the maximum take-off weight of the Airbus A500 is 280t. The battery alone would be 15.6 times the weight of a conventional, fully loaded plane.

Several alternatives to fossil fuels for powering aircraft are under investigation. The most promising is using hydrogen as fuel. The feasibility of powering aircraft with hydrogen has been demonstrated. Still, the technology is not yet mature enough for large commercial deployment and is unlikely to reach that stage in ten years. Clearly, we are unable to *decarbonise* aviation in the near future. How big a problem is this? EThe International Energy Agency estimates that during 2019 the world's total aviation fuel consumption was 331 Mt. With an energy density of aviation fuel of around 43 MJ/kg this makes $43\ 10^6\,\mathrm{J/kg} \cdot 331\ 10^9\mathrm{kg} = 14.233\ 10^{18}\mathrm{J}$ or $14.233\,\mathrm{EJ}/492.39\,\mathrm{EJ} \times 100 = 2,89\%$ of total world 2019 fossil fuel consumption. However, aviation contributes more to global warming than its share of CO_2 emissions. There are several causes for it; the main one is the jet engine contrails of water vapour. All in all, the aviation contribution to global warming is about 3.5% (Lee et al. 2021). Even so, aviation plays a relatively small role in global warming, and our current inability to decarbonise aviation poses no major obstacle in the quest to stop global warming. Interestingly, estimates indicate that 1% of the world population is responsible for 50% of the CO_2 emission from commercial aviation. Constraining frequent flyers could substantially reduce air travel emissions.

10

The Demise of Fossil Fuels

Fossil fuels are fossil hydrocarbon compounds. The same fossil hydrocarbons used as fuels also supply the raw materials for the petrochemical industry, which produces chemicals for a wide range of everyday goods, including plastics, synthetic fibers, medicines, and household products. Fossil hydrocarbons are not always bad. What is bad is burning them to obtain energy. Their use as raw materials in the chemical industry does not substantially contribute to global warming, provided there are no significant residual CO_2 emissions from the chemical reactions. Therefore, when examining the effect of the demise of fossil fuels, any reference to fossil fuels will strictly relate to the use of fossil hydrocarbons as fuels for energy generation.

The fossil fuel industry provides 87% (2024) of the total world energy supply and, in doing so, pervades the entire economy. It begins with prospecting for new coal, oil, and gas deposits. Then follow mining the coal or extracting oil and natural gas from these deposits, oil refining, the transport and distribution of all fossil fuel products, and finally,

J. Sitte, *Global Warming: Facts, Myths and Remedies*,
https://doi.org/10.1007/978-3-032-20862-0_10

retailing and marketing. Additionally, there are the manufacturing, installation, and maintenance of machinery driven by fossil fuels, including cars, trucks, electricity generators, airplanes, agricultural machinery, food processing equipment, and heating and cooling equipment.

In this chapter, we estimate the fossil fuel industry's contribution to the world economy, with the purpose of gauging the possible consequences of its demise on the world economy as a whole, for countries that depend heavily on the income from exporting fossil fuels, and for those where the cost of importing fossil fuels weighs heavily on their economy.

10.1 Fossil Fuel Industry Contribution to World GDP

The demise of fossil fuels means that the economic activity around their supply will gradually wind down and disappear. Initially, this loss will be replaced by the manufacture and installation of PV solar and wind energy equipment, along with the necessary solar energy storage. The industry for fossil-fuel-driven machinery may transform into an industry for the same kind of machinery powered instead by electricity. The key question is whether new economic activity in renewables will fully compensate for the decline in the fossil fuel supply chain.

Despite the world economy's heavy reliance on fossil fuels, it is hard to assign a monetary value to the industry's contribution. The industry's broad ramifications make it difficult to track the global fossil fuel value chain. Nevertheless, we will endeavour to obtain a rough estimate of the fossil fuel industry's contribution to global GDP.

Table 10.1 Estimated raw fossil fuel wholesale value for 2023

Type	Consumption	Average price (USD)	Value (trillion USD)
Coal	9095.7 Mt	100/tonne	0.91
Crude oil	$35.58\ 10^9$ barrels	80/barrel	2.93
Natural gas	$4.01\ 10^{12}\ m^3$	12/mmBtu	1.7
All fossil fuels			5.54

As a first step, let us estimate the yearly value of raw fossil fuel consumption by multiplying the raw consumption volumes by an estimated average wholesale price. We refer to the 2024 EI Statistical Review of World Energy, which provides data on raw (unprocessed) coal, oil, and gas consumption, as well as their prices and fluctuations over the year. What makes the calculation somewhat tedious is that crude oil, gas, and coal are not traded in the same units used to report consumption data, but rather in barrels for crude oil, British Thermal Units (BTU) for gas, and tonnes for coal. Thus, we need to do the corresponding conversions. The final hurdle is that all the fossil fuels come in several varieties (qualities) that command different prices. I have picked a plausible average price. Table 10.1 shows the results of the calculations.

The calculation indicates that trade in raw fossil fuels was worth around USD 5.5 trillion in 2023. However, there is more value added to fossil fuel products until they reach the consumers. To estimate the total contribution of the fossil fuel industry to world GDP, we need to add the value created along the value chain for all fossil fuel products. It is very laborious to track the various volumes and consumer prices of fossil fuel products across their entire value chains. A simple way to get a rough estimate of the value added to crude oil along the value chain is to compare the retail price of electricity with the bulk price of crude oil. The reason for choosing the electricity retail price is that electricity prices are less subject to widely varying taxation and subsidies around the world than, for example, petrol retail prices.

Spot prices for crude oil fluctuate considerably. Nevertheless, an average price of 80 USD/barrel for 2023 may not be too far off the mark. Looking at domestic electricity retail prices worldwide, the average appears to be somewhere around 0.20–0.25 USD/kWh. Let's work with an average price of 0.25 USD/kWh. By subtracting the cost of the crude oil consumed to generate 1 kWh from the retail price of electricity, we obtain the value added to the fossil fuel by electricity generation. Crude oil prices are quoted in US dollars per barrel, while electricity is charged in dollars per kWh. Therefore, we need to perform calculations to determine the cost of crude oil fuel required to produce 1 kWh of end-user electric energy. If we know how many kWh of electricity we get per barrel of crude oil, then we can obtain the fossil fuel cost of the electricity.

The *tonne of oil equivalent (toe)* energy, mentioned earlier, of 41.87 $GJ/tonne$ tells us the energy per unit mass of crude oil. However, the barrel is a measure of volume. To find the mass of a barrel, we need to multiply the volume of the barrel by the density. One barrel is 159 l. Crude oil has a range of densities (kg per litre). Let us use 0.830 kg/l as a reasonable estimate for an average. Thus, the mass of a barrel is

$$159\frac{\mathrm{l}}{barrel} \cdot 0.83\frac{\mathrm{kg}}{\mathrm{l}} = 131.97\frac{\mathrm{kg}}{barrel} = 0.132\frac{tonne}{barrel} \tag{10.1}$$

We get the energy in a barrel by multiplying the mass by the *toe*

$$0.132\frac{tonne}{barrel} \cdot 41.87\frac{\mathrm{GJ}}{tonne} = 5.527\frac{\mathrm{GJ}}{barrel} \tag{10.2}$$

The last step is to convert the Joules to kWh. One kWh is 3.6 MJ, hence we get

$$\frac{5.527\frac{\mathrm{GJ}}{barrel}}{3.6\frac{\mathrm{MJ}}{\mathrm{kWh}}} = 1535\frac{\mathrm{kWh}}{barrel} \tag{10.3}$$

Because fossil fuel-fed electric power plants operate with an efficiency of 40%, we only obtain 0.4 of the fossil fuel energy input as end-user electricity output.

$$1535\frac{\mathrm{kWh}}{barrel} \cdot 0.4 = 614\frac{\mathrm{kWh}}{barrel} \tag{10.4}$$

Finally, by dividing the price of a barrel by the kWh of electricity it can produce, we obtain the cost of the fossil fuel needed to generate 1 kWh of end-user electricity.

$$\frac{80\frac{\mathrm{USD}}{barrel}}{614\frac{\mathrm{kWh}}{barrel}} = 0.13\frac{\mathrm{USD}}{\mathrm{kWh}} \tag{10.5}$$

The ratio of the electric energy price to the fossil fuel input cost is

$$\frac{0.25\ \frac{\mathrm{USD}}{\mathrm{kWh}}}{0.13(\frac{\mathrm{USD}}{\mathrm{kWh}})} = 1.92 \tag{10.6}$$

meaning the consumer pays almost twice the cost of the fossil fuel used to generate the electricity. When we apply this factor to estimate the value added to the annual fossil fuel consumption of USD 5.54 trillion, we get USD 10.6 trillion. In terms of the world GDP for 2023 (107 trillion USD), it represents 10% of world GDP. Thus, we arrive at our estimate that the fossil fuel industry contributes approximately USD 10.6 trillion, or 10%, to world GDP.

This means consumers pay almost twice the cost of the fossil fuels used to generate electricity. Applying this factor to the annual fossil fuel consumption of USD 5.54 trillion results in USD 10.6 trillion. This amount is about 10% of the 2023 world GDP (USD 107 trillion). Therefore, the fossil fuel industry contributes approximately USD 10.6 trillion, or 10%, to world GDP.

How does the contribution of the fossil fuel industry compare with the carbon tax required for replacing the fossil fuel industry with solar electricity? First of all, the fossil fuel industry will not fully disappear until the transition to solar electricity is complete. If it takes the 20 years we aim for, it will decrease by 1/20 of USD 10.6 each year. With a carbon price of 85 USD/tonne calculated in Chap. 9 and a yearly emission of around 36 Gt of CO_2, the carbon tax will generate 3 trillion USD (2.8% of GDP) available in the first year for investment in the solar industry. This is six times the decrease in economic activity in the fossil fuel supply industry. Even if we grossly underestimated the value of the fossil fuel supply industry, the numbers indicate that the economic activity resulting from the transition to an all-electric economy will outweigh the loss of economic activity from the fossil fuel supply sector by a significant margin. Even more so if we raise the carbon price to also cover the cost of building electric energy storage infrastructure.

10.2 Geopolitical Consequences

The transition to an all-electric, solar-powered economy will make most countries energy self-sufficient. This shift will result in significant changes in their political and economic dependencies. Exporting countries will lose their clients as the demand for fossil fuels subsides. While the rise in the solar industry will more than compensate for the loss of

Table 10.2 Energy consumption from fossil fuels 2023 (EJ)

Country	Fossil fuels	Renewable energy	Percentage fossil of total
USA	75.97	18.31	0.806
Australia	5.12	0.90	0.851
Germany	8.56	2.84	0.751
China	139.24	31.49	0.816
India	34.80	4.20	0.892
World	504.83	91.56	0.815

economic activity from the fossil fuel industry, there will be large regional differences. Although predicting future economic development is often a futile exercise, in this chapter we will use what we have learnt about global warming, energy consumption, and wealth to identify broad trends and boundaries for the transition to a CO_2- emission-free economy.

10.2.1 The Global Fossil Fuel Market

With fossil fuels providing 87% of the total energy supply, the fossil fuel trade impacts every person's energy supply. Table 10.2 shows that in 2023, industrialised economies and populous countries alike, continue to heavily rely on fossil fuels for their energy needs.

Most countries must buy a large proportion of their crude oil consumption from the small number of producer countries listed in Table 10.3. The end of fossil fuel trading will have very different implications for each of these economies. The largest producer in 2023 was the USA, followed by Russia and Saudi Arabia, with China being the biggest importer.

The demise of fossil fuels will drastically reduce the income of the large exporters. Countries that heavily depend on fossil fuel sales urgently need to reduce their reliance on

Table 10.3 Crude oil production and consumption 2023 in kb/day (EI)

Country	Production	Consumption	Difference
USA	12,927	18984	−6057
Russia	10,554	3635	6919
Saudi Arabia	9609	4052	5548
Canada	4935	2351	2584
Iraq	4271	875	1920
China	4198	16577	−12379
Iran	3940	1817	2123
Brazil	3402	2567	835
UAE	3251	1139	2112
Kuwait	2590	411	2179
Kazakhstan	1891	349	1842
Mexico	1875	1962	−87
Norway	1818	213	1605
Qatar	1310	370	940
Venezuela	809	396	413

them. Some major oil producers have large, unused open spaces, mostly deserts, with high levels of solar irradiation. It has been suggested that these countries transform into solar energy suppliers. However, this idea has a fundamental flaw. In a solar energy world economy, most countries will be energy self-sufficient or nearly so and will not need to buy energy from elsewhere. Inequality in the distribution of energy resources, common in the fossil-fuel era, will be greatly reduced. Large fossil fuel exporters who consider becoming solar exporters will have a much smaller market. Their economies will likely decline sharply. The loss of political relevance of some of the previously dominant fossil fuel producers will have profound consequences for the existing world order. Large producers will no longer be able to influence energy prices by manipulating supply.

In the next sections, we will examine the relative economic weight of fossil fuel trade in some exporting and importing economies, as well as the potential changes that a carbon tax and the consequent reduction in fossil fuel trade may bring.

10.2.2 USA

The USA, despite being the largest producer, remains a large net importer of crude oil. The total fossil fuel consumption (oil, gas, and coal) of the USA in 2023 was 76.6 EJ, producing emissions of 4.64Gt of CO_2. A carbon price of 85 USD/tonne would make USD 0.394 trillion available for investment in PV solar, which is 1.42% of the 2023 USA GDP. The carbon tax would represent a small fraction of the country's GDP and would not disrupt the economy. Replacing fossil fuel consumption with solar electricity would reduce the USA's fossil fuel imports, depriving its current suppliers of income. About half of the USA's crude oil imports come from Canada, followed by Mexico, Saudi Arabia, Iraq, and Brazil. Thus, the country most affected economically would be Canada. The rest of the USA's fossil fuel imports are spread across many countries, so the vanishing of the USA market alone would not be critical to their economies. Furthermore, in 2019, the USA became a net energy exporter due to coal and natural gas exports. Interestingly, the USA is in a position to become self-sufficient in energy without transitioning to renewable energy by replacing its energy imports with its own production that it can no longer export.

10.2.3 Russia

After the USA, Russia was the second-largest crude oil producer and the largest exporter in 2023. Oil, gas, and coal accounted for around 60% of Russia's exports in 2023, equivalent to about 16% of its GDP. The loss of its oil and gas exports due to the decline of fossil fuels would be a substantial loss of foreign income; however, its diversified economy may mitigate the impact, preventing a drastic economic downturn. Russia's fossil fuel consumption in 2023 was 31.29 EJ with corresponding CO_2 emissions of 1614.7 Mt. At a carbon price of 85 USD, the initial carbon tax would be 137.25 billion USD, which is 6.8% of its 2.021 trillion USD GDP. The carbon tax would represent a redirection of GDP almost 5 times larger than in the USA, but would still be affordable. Russia can afford to reduce its CO_2 emissions at the rate required. Russia's emissions in 2023 accounted for 4.5% of global emissions, which is still only one-third of the USA's.

10.2.4 Saudi Arabia

Saudi Arabia is the third-largest producer of crude oil and the second-largest exporter, nearly on par with Russia. Around 90% of Saudi Arabia's export earnings come from fossil fuels, and the fossil fuel industry accounts for around 42% of the country's GDP. For Saudi Arabia, unlike Russia, the demise of fossil fuels poses a severe threat to its economy, which is heavily dependent on oil export income.

Saudi Arabia's fossil fuel consumption in 2023 was 11.54 EJ, and its CO_2 emissions were 620.4 Mt. A carbon price of USD 85 would result in USD 52.7 billion. That is 4.9% of its USD 1.068 trillion (2023) GDP. Thus, the transition to an electric economy would be more affordable for Saudi Arabia than for Russia. Despite the decline in global fossil fuel consumption and, consequently, its export income, the country can still afford to invest the required funds to replace its own fossil fuel consumption. With its proximity to the northern economies of Europe and Asia and its abundant sunshine, Saudi Arabia may find a market for exporting solar electricity to partially replace its oil and gas exports. Saudi Arabia would have to invest substantially more than the carbon tax would require in the infrastructure needed to become a major exporter of solar electricity. Saudi Arabia has invested part of its past fossil fuel income in the Public Investment Fund (PIF), holding over 600 billion USD in worldwide investments. It could sell these assets to fund its transition to becoming a solar energy supplier. Despite this, the demise of fossil fuels could cripple Saudi Arabia's economy.

10.2.5 Bangladesh

Turning to countries with low per capita energy consumption and hence low GDP, Bangladesh is the most populous country endangered by global warming through sea-level rise. Its CO_2 emissions from energy consumption totaled 118.2 Mt in 2023, accounting for 0.34% of global emissions. Bangladesh, like many other smaller economies around the world, has little influence on global warming, being completely at the mercy of the big CO_2 polluters. Its motivation for abandoning fossil fuels can only come from economic considerations and, perhaps, a sense of moral duty to contribute its share to halting global warming. The economic incentives are strong. Bangladesh imports 50.2% of its energy. Renewable energy production is very small. Per capita energy consumption in 2023 was 10.6 GJ, which is 14% of the world's mean per capita consumption. For Bangladesh, the path to prosperity, starting at a very low level, is to increase solar energy production. Bangladesh has significant natural gas reserves, but installing solar power generation is likely to be cheaper than developing its gas fields. At 85 USD/tonne, the carbon tax for Bangladesh's 2023 emissions would be USD 10 billion or 2.3% of its USD 437.4 billion GDP. Thus, Bangladesh may afford to transition to an all-electric economy without external assistance. The demise of fossil fuels may be a strong stimulant for Bangladesh's socio-economic development and an opportunity to raise per capita energy consumption to at least 1000 kWh (36 GJ), set as a goal for a*dignified and productive life* by the Energy for Growth Hub.

Bangladesh is a densely populated country, and we may ask whether it has the space for solar panels to provide its population with the world's mean per capita energy? If per capita energy consumption were to reach at least the 1000 kWh (36 GJ) per capita, stipulated or dignified and productive life, then it would need an installed effective PV generating capacity of

$$\frac{36\,\text{GJ}}{3600 * 24 * 365\,\text{s}} = 1.1416\,\text{kW} \tag{10.7}$$

per capita. For a population of 171.5 million, that translates to 0.196 TW, which requires 0.196 TW/0.193 = 1.016 TWp. At 180 Wp/m^2 this gives 5642 km^2, which is 3.8% of the area of Bangladesh. This number is less than the total estimated built-up area of 8300 km^2 in Bangladesh. There seems to be enough space, beyond available rooftop space, to replace its fossil fuel consumption and to increase per capita energy consumption to reach the target for a *dignified and productive life*!

10.2.6 India

Nearby, Bangladesh's neighbour, India, is, after China and the USA, the third-largest emitter of CO_2, with 2.814 Gt in 2023. The yearly carbon cost would be USD 239 billion, which amounts to 6.7% of its USD 3.568 trillion GDP (2023). For India, it would be much harder than for Bangladesh to pay for the transition to an all-electric economy. From its fossil fuel consumption of 39.02 EJ (2023), it imports about 1/3 of that. Furthermore, 63% of India's fossil fuel consumption is coal, of which it imports about one quarter. The remainder is mostly oil, of which it imports around 88%. We can get a rough estimate of the cost of India's fossil fuel imports by converting one-third of its energy consumption from Joules to tonnes of crude oil by dividing by the *toe*, then multiplying by the cost of a tonne of oil. At USD 80 per barrel, we get USD 188 billion , which is somewhat less than the initial carbon tax. In monetary terms, it is cheaper for India to keep importing than to levy a carbon tax, providing little short-term incentive for transitioning to solar electricity.

10.2.7 China

In 2023, China was the biggest emitter of CO_2, with 11218.4 Mt. The carbon cost of this emission would be 0.95 trillion USD. This is 5.36% of China's 2023 GDP of 17.79 trillion. Out of its 2023 fossil fuel consumption of 170.74 EJ it imported 44 EJ (26%), mostly oil and gas. Because of the sheer size of its economy, China is the world's largest importer of crude oil. China produces almost all the coal it consumes. An estimate of China's fossil fuel import cost is USD 0.824 trillion. As in India, the short-term monetary incentive for transitioning to an all-electric economy is low, and the impact on the trade balance is also modest. China's incentive to replace fossil fuel consumption with solar energy can come from environmental concerns about being the largest CO_2 emitter and from the potentially lower cost of renewable energy.

Table 10.4 2019 German electricity generation data

Consumption/generation	TWh	EJ
Electricity generation	612.4	2.205
Non-fossil fuel electricity	224.1	0.807
Fossil fuel electricity	388.3	1.398
Electricity generation from coal	171.2	0.616
Energy consumption from coal		2.3

Electricity generation is measured by gross output

10.2.8 The Bottom Line

From the analysis of the countries above, we see that China, the USA, Russia, and India, which are responsible for 58% of the world's emissions and account for 45% of the world's population, have no strong short-term economic incentive to transition to an all-electric economy.

10.3 An Example: Germany's Plan

In November 2021, the then-new German coalition government announced its energy transition policy. The two main goals for 2030 were:

1. Phase out the consumption of coal.
2. Raise renewable electricity generation to 80%.

Table 10.4 shows the situation of the German electricity generation in 2019 according to the 2020 BP Statistical Review of World Energy(British Petroleum, 2020). The electricity demand predicted for 2030 is 750 TWh (2.7 EJ). Of this, 80% (600 TWh) will have to come from renewable electricity generation. The remaining 20% (150 TWh) must come from oil and gas, not coal. To meet the goal, renewable electricity generation has to increase by 525.9 TWh (1.89

EJ) from 224.1 TWh (0.807 EJ) in 2019 to 750 TWh (2.7 EJ) in 2030. Let us assume that *renewable* refers to non-fossil fuel sources.

What would be the cost of increasing renewable energy generation to 1.89 EJ? For simplicity, let us assume that the entire increment is generated by PV solar electricity. We estimated earlier that the average daily generating capacity of crystalline PV solar panels is 16.67 MJ per kWp. Over a year, this is 16.67 MJ × 365 = 6.085 GJ per kWp. To generate 1.89 EJ (525.9 TWh) per year, Germany needs to install

$$\frac{1.89\ 10^{18}\,\mathrm{J}}{6.085\ 10^{9}\,\mathrm{J/kWp}} = 0.312\ 10^{9}\mathrm{kWp} \qquad (10.8)$$

until 2030. At a cost of USD 1.5 per Wp, that would be

$$1.5\frac{\mathrm{USD}}{\mathrm{Wp}} \cdot 0.312\ 10^{12}\mathrm{Wp} = 468\ 10^{9}\ \mathrm{USD} \qquad (10.9)$$

or 468/11 = 42.6 billion USD per year if spread evenly over the 11 years from 2019 to 2030. Compared to Germany's 2019 GDP of USD 3.957 trillion, the annual cost of the additional PV solar panel installation is 1.1% of that GDP. This amount seems to be quite affordable for the German economy.

Figure 10.1 shows what has happened to Germany's Plan so far. According to the 2024 Statistical Review of World Energy, Germany's total electricity generation was 513.7 TWh (1.85 EJ). Of this, non-fossil fuels accounted for 303.2 TWh (1.09 RJ), while coal still accounted for 127.8 TWh. In the four years since 2019, non-fossil fuel electricity production increased from 224.1 to 303.2 TWh, a rise of 79.1 TWh (0.285 EJ). With a steady increase, the rise should have been 191.2 TWh (0.688 EJ), not 79.1. Only two years after

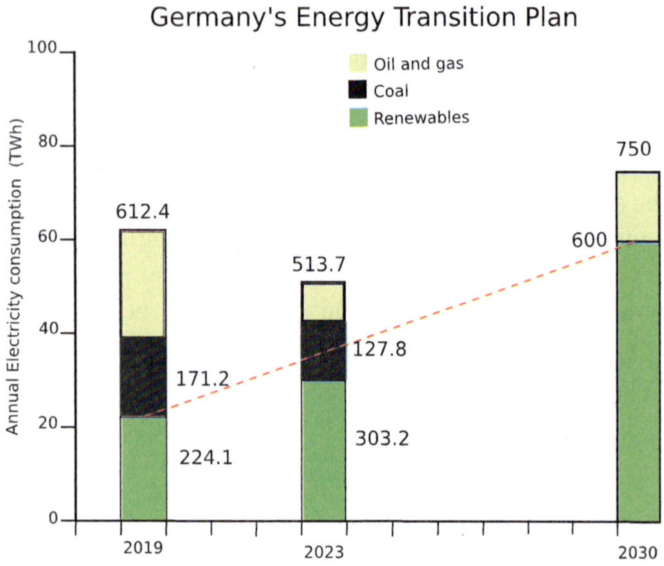

Fig. 10.1 Germany's Energy Transition Plan. Annual electricity consumption by fuel type (top: oil and gas, middle: coal, bottom: renewables). The dotted line shows linear progression toward the 2030 target of 80% renewables

the plan was announced, Germany had already fallen short by 112 TWh (0.4 EJ).

10.4 Demand for Critical Minerals for the All-Electric Economy

Although replacing the fossil fuel economy with an all-electric economy will, for many countries, end their reliance on a small group of fossil fuel producers, it introduces a new dependence on a different group of countries that possess the special minerals needed for solar electricity technology. This dependence has become quite extreme in the 2020s, where China has almost a total monopoly on the mining and refining of the rare-earth minerals so essential for the magnets in electric motors and wind turbine generators. As noted in Chapter 8, China is the lowest-cost manufacturer of PV solar panels, giving it a stranglehold on its production.

10.5 Global Warming and Population Growth

In a world dominated by fossil fuel energy, CO_2 emissions grow with the size of the human population. As the world population grows from 8 billion in 2022 to an estimated 10 billion in 2060. Assuming that the mean per capita energy consumption remains constant, the 20% increase in population extrapolates to a 20% increase in energy consumption. In 2023, for a large proportion of the world population, per capita energy consumption was still below what is today considered necessary for a *dignified and productive* life. Raising that part of the population out of energy poverty will increase overall energy consumption even without population growth. In turn, in an economy powered by solar energy, energy consumption can grow with the population without posing a threat. Although population growth will continue to be a limiting factor in the use of other resources, including the loss of animal habitats and reduced food production.

10.6 Consequences of Adaptation

Inaction to address global warming will inevitably lead to forced adaptation. Because rich nations have more resources to adapt to climate impacts, they will become relatively more attractive destinations for people living in poorer countries, where adaptation may not be possible. As a result, more people from areas rendered unlivable by climate warming may migrate to wealthier nations. This movement may further intensify existing social pressures related to migration. If fossil fuels are replaced with renewable energy, however, poorer economies could benefit from higher per capita energy consumption, thereby reducing poverty and lessening migratory pressures.

Epilogue

Thanks to its proclivity for record-keeping, humankind spotted global warming early. And, thanks to our inquisitiveness, we were able to relate the rise in Mean Global Surface Temperature to the sudden increase in atmospheric CO_2 concentration. In turn, we found the rapid expansion of fossil fuel burning to be the cause of the rise in CO_2, In the burning of fossil fuels, we found an easy way to obtain the energy needed to drive the ever-increasing affluence that the industrial revolution and the technical civilisation we built on it gifted to a fortunate fraction of humankind.

Thanks to the internet, all relevant data on human affluence, energy production and consumption, and their consequences for the human environment and the climate are available to everyone. Every one of us can make his/her own judgment about the predicament we find ourselves in.

Unfortunately, it is human nature to interpret events in ways that push aside threats and reinforce ideas and convictions deemed essential to protecting one's privileges and prosperity. Therefore, different views emerge, leading to

J. Sitte, *Global Warming: Facts, Myths and Remedies*,
https://doi.org/10.1007/978-3-032-20862-0

controversies and conflicts of interest that inhibit prompt, decisive action in the face of a threat.

As humans, we are often easy prey to manipulation in situations that we do not fully grasp. It may be due to indifference, lack of engagement, lack of knowledge, or simply feeling overwhelmed by the magnitude of the task before us. Leaders of our communities often resort to manipulation when their personal interests do not align with their communities' needs. This was and continues to be the case in the global warming debate.

To overcome the paralysis evident in acting to mitigate or stop global warming, it is essential that we share a common awareness of the threat, that the majority has a shared conviction of its significance, and that we can conceptualise effective actions to counter it.

The chapters of this little book contain the most pertinent data on the threat of global warming, along with guidelines for their analysis. If you had the motivation and perseverance to work your way through the chapters, you can be confident that you are well-equipped to reach your own well-grounded judgments on the facts, myths, and remedies to the global warming threat.

A

Baetz' Law

Baetz' law is easy to prove. The energy extracted by the turbine is the difference between the kinetic energy of the mass of air Δm flowing into the turbine and the kinetic energy of that mass of air coming out of the turbine in a time interval Δt.

$$
\begin{aligned}
E_e &= E_i - E_o = \frac{1}{2}\Delta m \cdot v_i^2 - \frac{1}{2}\Delta m \cdot v_o^2 \\
&= \frac{1}{2}\Delta m \cdot (v_i^2 - v_o^2)
\end{aligned}
\tag{A.1}
$$

J. Sitte, *Global Warming: Facts, Myths and Remedies*,
https://doi.org/10.1007/978-3-032-20862-0

Per unit time this is the power

$$\frac{E_e}{\Delta t} = \frac{E_i - E_o}{\Delta t} \tag{A.2}$$

$$P_e = P_i - P_o \tag{A.3}$$

$$= \frac{1}{2}\frac{\Delta m}{\Delta t} \cdot (v_i^2 - v_o^2) \tag{A.4}$$

$$= \frac{1}{2}\dot{m}(v_i^2 - v_o^2) \tag{A.5}$$

The extracted power depends on the incoming air speed v_i and the outgoing air speed v_o. Instead of looking at each of the speeds separately it is easier to look at the ratio v_o/v_i of outgoing speed to incoming speed. If the ratio is zero then all the energy has been extracted ($v_O = 0$). If the ratio is one ($v_o = v_i$ then no energy has been extracted.

The available power is the incoming power P_i. The fraction of the power extracted form the available power is

$$\frac{P_e}{P_i} = \frac{P_i - P_o}{P_i} \tag{A.6}$$

$$= \frac{\frac{1}{2}\dot{m}_e(v_i^2 - v_o^2))}{\frac{1}{2}\dot{i}\, v_i^2} \tag{A.7}$$

The important thing here is that the rate of mass flow $\dot{m}_e$, when energy is extracted from the air flow, is different from the incoming mass flow $\dot{m}_i$ through the area of the blades when the turbine is not operating. When the turbine is operating the air coming in at speed v_i is slowed down by the blades and leaves at a lower speed v_o. Let us look at what happens during a time interval Δt. At the start of the interval the air that will reach the plane of the blades at the end of the interval is the distance $d_{last} = v_i \cdot \Delta t$ away from the blades. It will move with speed v_i during the whole of the interval

Δt. All the air that is closer will pass through the plane of the blades before the end of the time interval Δt. The air that is closest to the blades at the beginning of the interval, will pass through the blades and move at a speed v_o during the time interval covering a distance $d_{first} = v_o \cdot \Delta t$ during the time interval. The air between the first and last to reach the plane of the blades will take less than time Δt to reach the blades and will move at the the lower speed for the rest of the time interval. We see that the distances that the air moves in the time interval Δt depends on how far the air is from the plane of the blades and varies between d_{first} and d_{last}. The average distance the air travels during the time interval Δt is $d_{mean} = \frac{1}{2}(d_{last} + d_{first})$. Dividing the distance by the time taken to travel we find that the air moves at a mean speed of $v_{mean} = \frac{1}{2}(v_i + v_o)$. The mass flow is $\dot{m}_e = \rho \cdot A \cdot v_{mean}$.

$$\frac{P_e}{P_i} = \frac{\rho \cdot A \cdot v_{mean}(v_i^2 - v_o^2))}{\rho \cdot A \cdot v_i^3} \tag{A.8}$$

$$= \frac{\frac{1}{2}(v_i + v_o) \cdot (v_i^2 - v_o^2)}{v_i^3} \tag{A.9}$$

$$= \frac{1}{2}(1 + \frac{v_o}{v_i})(1 - (\frac{v_o}{v_i})^2) \tag{A.10}$$

As shown in Fig. A.1 the fraction of extracted power as a function of the ratio v_o/v_i first increase with the ratio and then decreases, having a maximum at $v_o/v_i = 1/3$.

We can easily calculate the optimal ratio v_o/v_i by setting $x = \frac{v_o}{v_i}$ in Eq. (A.10) to get the equation

$$y = \frac{1}{2}(1 + x)(1 - x^2) = \frac{1}{2}(1 + x - x^2 - x^3) \tag{A.11}$$

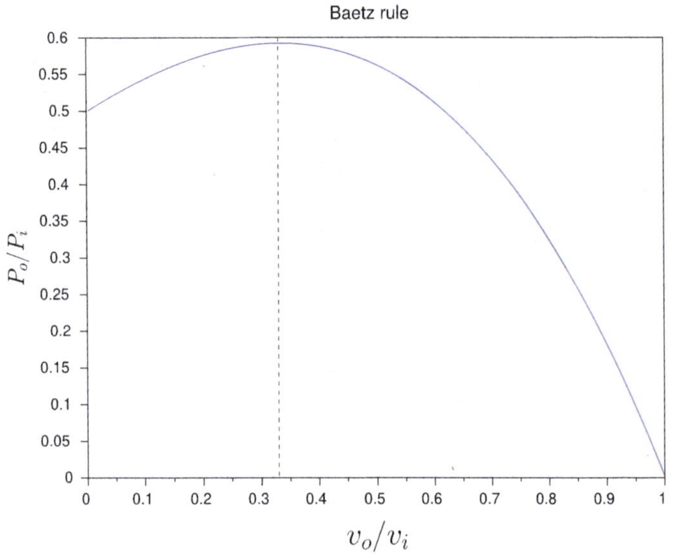

Fig. A.1 Wind turbine efficiency (P_e/P_i) as a function of the ratio of outgoing wind speed v_o to incoming wind speed v_i

To find the maximum we set the derivative to zero

$$\frac{dy}{dx} = \frac{1}{2}(1 - 2x - 3x^2) = 0 \tag{A.12}$$

and find the positive root of this quadratic equation. The result is $x_{max} = 1/3$, and $y_{max} = y(x_{max}) = y(1/3) = 16/27$.

References

Abbott, Derek. 2011. Is nuclear power globally scalable? *Proceedings of the IEEE* 99 (10): 1611–1617.

Arrhenius, Svante. 1986. On the influence of carbonic acid in the air upon the temperature of the ground. *Philosophical Magazine and Journal of Science, Series* 5 (41): 237–276.

Berndes, G., M. Hoogwijk, and R. van den Broekc. 2003. The contribution of biomass in the future global energy supply: a review of 17 studies. *Biomass and Bioenergy* 25 (D20): 1–28.

Berner, Robert A., and Kothavala Zavareth. 2001. Geocarb III: a revised model of atmospheric CO_2 over phanerozoic time. *American Journal of Science* 301 (2): 182–204.

Bernreuter, J. 2022. *Polysilicon manufacturers.* https://www.bernreuter.com/polysilicon/manufacturers/. Retrieved 9 June 2022.

Bowker, M. 2019. Methanol synthesis from CO_2 hydrogenation. *ChemCatChem* 11: 4238–4246.

J. Sitte, *Global Warming: Facts, Myths and Remedies*,
https://doi.org/10.1007/978-3-032-20862-0

British Petroleum. 2020. *BP statistical review of world energy.* Technical Report 69th Edition, British Petroleum.

Burkholder, J.B. (Lead Author), Ø. Hodnebrog, and V.L. Orkin. 2018. Summary of abundances, lifetimes, ozone depletion potentials (ODPs), radiative efficiencies (REs), global warming potentials (GWPs), and gobal temperature change potentials (GTPs). Chapter Appendix A of Scientific Assessment of Ozone Depletion: 2018. World Meteorological Organization.

Burzlaff, Marcus. 2017. *Aircraft fuel consumption—estimation and visualization.* Technical Report, Hamburg University of Applied Science I.

Chalamala, B. 2018. *Manufacturing of silicon materials for microelectronics and solar PV.* https://www.osti.gov/servlets/purl/1497235. Retrieved 13 June 2022.

Cheng, Chin-Hsien., and Simon A. T. Redfern. 2022. Impact of inter annual and multidecadal trends on methane-climate feedbacks and sensitivity. *Nature Communications* 13: 3592.

Coase, R. H. 1988. *The firm.* The Market and the Law: University of Chicago Press.

Prá, Dai, Léa. Beatriz, João Batista. Dias, and Amanda Gonçalves Kieling. 2015. Comparison between the energy required for production of PV module and the output energy throughout the product life time. *Journal of Energy and Power Engineering* 9: 592–597.

EI. 2024. *Energy institute statistical review of world energy.* Technical Report 73rd ed, Energy Institute.

Energy for Growth Hub. *The modern energy minimum: the case for a new global electricity consumption threshold.* https://energyforgrowth.org/wp-content/uploads/2019/01/FULL-Modern-Energy-Minimum-final-Jan2021.pdf. Retrieved 7 October 2023.

EPA. 2022. *Climate change indicators: atmospheric concentrations of greenhouse gases.* https://www.epa.gov/climate-indicators/climate-change-indicators-atmospheric-concentrations-greenhouse-gases. Retrieved 29 June 2022.

Farley, Peter. 2023. Buoyant behemoths. *IEEE Spectrum* 30–37.

Fleming, J. R. 1998. *Historical perspectives on climate change.* Oxford University Press.

Frankopan, Peter. 2023. *The earth transformed: an untold history.* Bloomsbury Publishing.

Friedlingstein, P., et al. 2020. Global carbon budget 2020. *Earth System Science Data* 12: 3269–3340.

Gates, Bill. 2022. *How to avoid a climate disaster.* Penguin Books.

Guo, J., D. Kubli, and P. Saner. 2021. *The economics of climate change: no action not an option.* Technical Report.

Gupta, Neeraj. 2016. A review on the inclusion of wind generation in power system studies. *Renewable and Sustainable Energy Reviews* 59 (06): 530–543.

Hansel, Rudolf A. et al. 1994. IRIS/Nimbus-4 level 1 radiance data V001. Retrieved 04 Sept 2025.

HRE4. 2017. *Heating and cooling, facts and figures.* https://heatroadmap.eu/wp-content/uploads/2019/03/Brochure_Heating-and-Cooling_web.pdf. Retrieved 5 Sept 2022.

Huang, Boyin, Xungang Yin, Russell S. Menne, Matthew J. and Vose, and Huai-Min Zhang. 2024. NOAA global surface temperature dataset (NOAAGlobalTemp) aravg.mon.land_ocean.90S.90N.v6.0.0.202412. https://doi.org/10.25921/rzxg-p717. Retrieved 6 Feb 2025.

IEA. 2021. *Key world energy statistics 2021.* Technical Report, International Energy Agency.

IEA. 2023. *Renewable energy market update—June 2023.* Technical Report, International Energy Agency.

Jacobs, R.P.W.M. et al. 1988. The greenhouse effect. Technical Report HSE 88–001, Shell Internationale Petroleum Maatschappij B.V., Health, Safety and Environment Division (HSE).

Jester, Theresa L. 2002. Crystalline silicon manufacturing progress. *Progress in Photovoltaics: Research and Applications* 10 (2): 99–106.

Jouzel, J., et al. 2007. *EPICA dome C ice core 800KYr deuterium data and temperature estimates.* Technical Report.

Judd, Emily, Jessica Tierney, Daniel Lunt, Isabel Montanez, Brian Huber, Scott Wing, and Paul Valdes. 2024. A 485-million-year history of earth's surface temperature. *Science (New York, N.Y.)* 385: eadk3705.

Kaya, Yoiichi, and Keiichi Yokobori. 1997. Environment, energy, economy and sustainable development. In *Environment, energy and economy: Strategies for sustainability, proceedings of the Tokyo conference on global environment, energy, and economic development, October 1993*, edited by Yoiichi Kaya and Keiichi Yokobori. United Nations University Press.

Keeling, Charles D. 1960. The concentration and isotopic abundances of carbon dioxide in the atmosphere. *Tellus* 12 (2): 200–203.

Kill, Jutta, Saskia Ozinga, Steven Pavett, and Richard Wainwright. *Trading carbon, how it works and why it is controversial.* https://www.fern.org/publications-insight/trading-carbon-how-it-works-and-why-it-is-controversial-651/. Accessed Nov 2024.

Kroposki, Benjamin, and Andy Hoke. 2024. A path to 100% renewable energy. *IEEE Spectrum* 61 (5): 50–57.

Kühne, Kjell, Nils Bartsch, Ryan Driskell Tate, Julia Higson, and André Habet. 2022. Carbon bombs—mapping key fossil fuel projects. *Energy Policy* 166: 112950.

Lee, D. S., D. W. Fahey, A. Skowron, M. R. Allen, U. Burkhardt, Q. Chen, S. J. Doherty, S. Freeman, P. M. Forster, J. Fuglestvedt, A. Gettelman, R. R. De León, L. L. Lim, M. T. Lund, R. J. Millar, B. Owen, J. E. Penner, G. Pitari, M. J. Prather, R. Sausen, and L. J. Wilcox. 2021. The contribution of global aviation to anthropogenic climate forcing for 2000 to 2018. *Atmospheric Environment* 244: 117834.

Lindsey, R. 2012. *Climate and earth's energy budget.* https://earthobservatory.nasa.gov/features/EnergyBalance. Retrieved 12 Sept 2021.

Masson, Gaëtan, Melodie De l'Epine, and Izumi Kaizuka. 2024. *Trends in photovoltaic applications 2024.* Technical Report EA PVPS T1–43:2024, International Energy Agency.

Masson-Delmotte, V. 2021. *Climate change 2021: the physical science basis. Contribution of working group I to the sixth assessment report of the intergovernmental panel on climate change.* Cambridge University Press.

Maurits, J. E. A. 2014. *Silicon production.* Chapter 2.6 of Treatise on Process Metallurgy, vol. 3, 919–948. Elsevier.

Merz, Joseph J., et al. 2023. World scientists' warning: the behavioural crisis driving ecological overshooty. *Science Progress* 106 (3): 1–22.

Moore, Patrick. 2021. *Fake invisible catastrophes and threats of doom, independently published.*

Myhre, G., E. Highwood, K. Shine, and F. Stordal. 1998. New estimates of radiative forcing due to well mixed greenhouse gases. *Geophysical Research Letters* (14): 2715–2718.

Ncei. 2022. *Billion-dollar weather and climate disasters.* https://www.ncei.noaa.gov/access/billions/. Retrieved 13 June 2022.

NOAA. 2021. *The NOAA annual greenhouse gas index (AGGI).* https://gml.noaa.gov/aggi/aggi.html. Retrieved 29 Dec 2021.

Olivier, J. G. J., and J. A. H. W. Peters. 2020. *Trends in global CO_2 and total greenhouse gas emissions: 2020 Report.* Technical Report.

Peiseler, Leopold et al. 2024. Carbon footprint distributions of lithium-ion batteries and their materials. *Nature Communications* (10301).

Petroni, Agostino. 2023. *The Italian farmers saving an ancient fruit with solar power.* https://www.bbc.com/future/article/20230424-how-agrivoltaics-helped-save-italys-citron-and-lemon-trees. Retrieved 18 Sept 2023.

Qiao, Qinyu, et al. 2017. Comparative study on life cycle CO_2 emission from the production of electric and conventional vehicles in China. *Energy Procedia* 105: 3584–3595.

Ramasamy, Vignesh, Jarett Zuboy, Michael Woodhouse, Eric O'Shaughnessy, David, Jal Desai Feldman, Andy Walker, Robert Margolis, and Paul Basore. 2023. 2023. *U.S. solar photovoltaic system and energy storage cost benchmarks, with minimum sustainable price analysis: Q1 2023.* Technical Report NREL/TP-7A40-87303, National Renewable Energy Laboratory.

Ravishankara, A. R., et al. 2021. *Global methane assessment: benefits and costs of mitigating methane emissions.* United Nations Environment Programme and Climate and Clean Air Coalition: Technical Report.

Riebeek, Holli. 2011. *The carbon cycle.* https://earthobservatory.nasa.gov/features/CarbonCycle. Retrieved 15 May 2023.

Royer, Dana L., Robert A. Berner, Isabel P. Montanez, Neil J. Tabor, and David J. Beerling. 2024. CO_2 as a primary driver of phanerozoic climate. *GSA Today* 14 (3): 4–10.

Sanchez-Bastardo, N., R. Schlögl, and H. Ruland. 2021. Methane pyrolysis for zero-emission hydrogen production: a potential bridge technology from fossil fuels to a renewable and sustainable hydrogen economy. *Industrial and Engineering Chemistry Research* 60: 11855–11881.

Scotese, Christopher R., Haijun Song, Benjamin J. W. Mills, and Douwe G. van der Meer. 2021. Phanerozoic paleotemperatures: the earth's changing climate during the last 540 million years. *Earth-Science Reviews* 215.

Shaviv, Nir J., Henrik Svensmark, and Jan Veizer. 2023. The phanerozoic climate. *Annals of the New York Academy of Sciences* 1519 (1): 7–19.

Smil, Vaclav. 2017. *Energy transitions: global and national perspectives.* Praeger.

Steinhilber, F., J. Beer, and C. Fröhlich. 2009. Total solar irradiance during the Holocene. *Geophysical Research Letters* 36(19).

U.S. Geological Survey. 2023. *Mineral commodity summaries 2023.* https://doi.org/10.3133/mcs2023. Retrieved 12 May 2025.

Index

A

Absorption lines, 75
Air flow, 185
Air travel, 229
  emissions, 231
Albedo, 181
Ammonia, 144
Ampere-hour, 193, 197
Anders, 29
Anode, 194
Antarctic
  ice core, 4, 65
  temperature record, 66
Anthacite, 102
Argons, 42
Arrhenius, 1, 31, 72
  calculations, 73
Artic, 7
Asimov, 203
Atmosphere, 32
  absorption, 39
  mass, **20**, 184
  radiation, 41
  reflection, 39
Atmospheric window, 53
Atom, 97
Atomic
  mass, 98
  number, **98**
  weight, **98**
Atomic weight, 98
Avogadro's number, **19**, 99

B

Baetz's law, 189, 253
Barrel (unit), 141, 171

J. Sitte, *Global Warming: Facts, Myths and Remedies*,
https://doi.org/10.1007/978-3-032-20862-0

Battery, 153, **193**
current, 193
current flow, 194
Li-ion, 195
21700 cell, 196
manufacturing emissions, 200
stored energy, 194
thermal, 202
Battery-Electric Vehicle (BEV), 138
Biofuels, 154
traditional, **113**
Biomass, 151, 154
Bituminous coal, 102
Bricks, 202

C

Carbon
budget, 10
credit, 207
cycle, 80, 92
debt, 8
fee, 206
footprint, xiii, 200
offset, 208
tax, 209, 214
revenue, 214
Carbon Capture and Sequestration (CCS), 143, **149**
Carbon Dioxide (CO_2), 1
concentration, 4, 73
emission, 13, 58
budget, 79
international trade, 129
rights, 207
equivalent, 86
geological record, 80
measurement, 24
outgassing, 67
pre-industrial level, 73, 78, 83
reconstruction, 68
removal, 80
solubility, 65
Carbonic acid, 72
Carbon monoxide, 142, 173
Cathode, 194
materials, 196
Causal relation, 4, 64
Chemical formula, **97**
Climate, 6
models, 7, **58**
scenarios, 58
scientists, viii
Climate change, viii, **4**
critics, viii
prediction, 59
Clouds, 40, 46, 181, 203
Coal, 102, 142
combustion, 102
Combustion reaction, **96**
methane, 96
Conferences of the Parties (COP), xiv
Correlation, 4, 64, 124
Crude oil, 102
price, 171, 236
Czochralski method, 173, 176

D
Deuterium, 98

E
Earth surface
  absorption, 40
  area, 20, 38
  reflection, 40
Efficiency, **27**, 156
  electric motor, 138
  factor, 113, 167, 188
  ICE, 138
Efficiency factor, 210
Electric Vehicle (EV), **xiii**, 137, 220
  emissions, 221
Electrode, 193
Electrolysis, 137
Electron, 97
  mass, 97
Emission Trading Scheme (ETS), **207**, 209, 213

Energy, **x**, **26**
  balance, 10, 30
  chemical, 153
  conservation, 116
  consumption, 110, **111**, 113, 115, 120
    direct, 118
    indirect, 118
    inequality, 125
    international trade, 127
    internet, 120
    per-capita, 118
    target, 129
  density, 103
    ammonia, 144
    battery, 193
    hydrogen, 135
    jet fuel, 230
    Li-ion cell, 196
    methanol, 146
  kinetic, 26, 184
  mechanical, 116
  potential, 169, 201
  primary, 112, 114
  radiative, 26
  renewable, **113**
  solar, 10
    investment, 170
  specific, 103
    ammonia, 144
    battery, 193
    hydrogen, 134
    Li-ion cell, 197
    methanol, 146
    woody biomass, 154
  supply, **113**, 116, 117
    per-capita, 117
  thermal, 26, 31, 202
  units, 27
  waste, 130
  wind, 184
Energy flux, **36**
  solar, 36, 163
  wind, 190
Energy Institute, 112
Energy poverty, 130
Eon, 68
Equilibrium Climate Sensitivity (ECS), **78**
Explain, 14, 16

F

Farenheit, 59
Ferrosilicon, 180
Foraminifera, 68
Fossil fuel, **5**, 95, 96, 107
  consumption rate, 210
  energy generation, 10
  expenditure, 170
  exporters, 240
  market, 239
  producers, 1, 240
Fourier, 31, 72
Fuel Cell EV (FCEV), 138

G

Geothermal energy, 112, 123, 153
GFM, **153**
Global Mean Surface Temperature (GMST), ix, 2, 31, 60, 63
  100-year average, 60
  dataset, 2
  increase, 73
  NOAA data set, 60, 62, 76, 79
  reconstruction, 68
Global Warming Potential (GWP), **87**
  methane, 88, 89
Graphite, **102**, 195
Greenhouse effect, 30, 62
Greenhouse gases, ix, 42
Gross Domestic Product (GDP), 121
  fossil fuel contribution, 234

H

Happiness score, 125
Heat, 26
  capacity, 202
Hybrid Plug-in Electric Vehicle (HPEV), **xiii**, 227
Hydrocarbon compounds, 101, 233
Hydrogen, 134
Hydrogen economy, 134

I

Industrial revolution, 63, 93
Infrared radiation, 33
Infrared spectrum, 43
  CO_2, 46
  Earth, 52
  methane, 49
  ozone, 51
  water, 46
Input equivalent, 112
Intergovernmental Panel on Climate Change (IPCC), viii, 55, 62, 77
  Assessment Report, 2, 58

Internal Combustion Engine (ICE), **xiii**, 27, 130, 134, 137, 156, 197, 219, 220
Internal Combustion Engine (ICE) core, 2, 59

International Energy Agency (IEA), **89**, 120, 156, 177, 178, 202, 216, 218
International System of Units, 27
Interstellar space, 31
Isotope, **98**

J

Jet stream, 185
Joule, 27

K

Keeling curve, 4, 63, 76
Kyoto Protocol, 207

L

Light, 26
Lignite, 102
Liquefaction
  hydrogen, 135
  methane, 139
Lithium, 195
  ferro-phosphate, 196
  reserves, 199
Little ice age, 72

M

Maunder minimum, 72
Measurement, 17
Methane, 48, 87, **96**
  concentration, 88
Microwaves, 203
Milankovitch cycles, 66
Model, 14
  quantitative, 14
  radiative transfer, 75
Mole, 19, **99**, 101
Molecular concentration, 76, 88
Molecular weight, 24, **98**
Moon, 29

N

Natural gas
  textbf, 96
Neutron, 97
  mass, 97
Nimbus 4, 52
Nitrogen, 42, 144
Nucleus, 97

O

Occam's razor, **15**, 23
Ocean surface
  area, 85
Output
  effective, 167
Output nominal, 167
Oxygen, **24**, 42, 51, 96, 134, 137, 195
Ozone, **51**

P

Paris Accord, 62, 79
Periodic table, 97
Perovskite, 164
Phanerozoic, 68

Photovoltaic (PV), **152**
Photovoltaic solar
  cells, 152, **178**
    crystalline silicon, 163, **172**
    light absorption, 163
    tandem, 164
    thin-film, **172**
  energy, 162
  panels, **174**
    conversion efficiency, 163, 164
Polysilicon, 173
  solar-grade, 173, 176
    supply, 180
Population growth, 117, 249
Pouillet, 31
Power
  consumption rate, 210
  continuous, 167
  effective, 167
  enduser, 210
  peak, 166
Prediction, 14, 32
Prosperity, 7
Proton, 97
  mass, 97
Pumped hydroelectricity, 140, **153**, 200

Q

Quartz, 172

R

Radiation
  balance, 72
  electromagnetic, 31
  thermal, 31
Radiation flux, 54
Radiative forcing, 54, 75, 86
  NOAA calculations, 84
Renewable electricity, 140

S

Science, 15
Scientific
  fact, 17
  method, 15
  notation, 19
Sediment layers, 59, 68
Shell Petroleum, 1
Silica, 172
Silicon, 172
  crystalline, 172
  metallurgical, 173
  oxide, 172
  polycristalline, 173
  production, 180
Solar
  cells, 162
  farm in space, 203
  heating, 162
  panel
    energy cost, 175
    lifetime, **216**
    manufacture, 172
    production capacity, 177, 214
Solar irradiance, 65, 69, 70
Specific heat, 85
Spodumene, 195

Standard conditions, **19**, 135
Statistical Review of World Energy, 80, 106, 107, **112**
Steam reformation, 142
Sunshine hours, 165
Sunspot cycle, 36, 70

T

Temperature, viii
  anomaly, 60
  measure, 2, 59
  pre-industrial level, 62
  record, 5, 59
  surface, 60
Thermal conduction, 31
Thermal solar
  energy, 162
  panels, 152
Thermometer, 59
Tonne of oil equivalent (toe), **102**, 236
Total energy supply, **113**
Total Solar Irradiance (TSI), **69**
Tritium, 98
Troposphere, 185
Tyndall, 31, 72

U

Understanding, 14
UN Environment Program (UNEP), **90**
US National Oceanic and Atmospheric Administration (NOAA), 2, 55, 60, 84

W

Wafer, **173**, 178
Warming, 1, 72
  trend, 6, 61
Water vapour, 45, 52, 134
Watt, 28
Weather, 93
  extreme, 6
  forecasts, 59
  pattern, 7
Wind
  energy, 183
  farms, 191
  turbine, **183**
    blades, 185
    land area, 191
Windmill, 185
Wood mass, 110
World population, 125

Y

Yoichi Kaya equation, 123

9783032208613